Zeitschrift für Meteorologie

Band 27 · 1977

Im Auftrage der
Meteorologischen Gesellschaft
der Deutschen
Demokratischen Republik

herausgegeben von
K. Bernhardt, Berlin,
E. A. Lauter, Berlin und
H. Pleiß, Dresden

Akademie-Verlag · Berlin

Schriftleitung: Prof. Dr. *K. Bernhardt*, Berlin, Prof. Dr. *E. A. Lauter*, Berlin, und Prof. Dr. *H. Pleiß*, Dresden · Verantwortlicher Redakteur: Dr. *R. Haake*, Potsdam
Verlag: Akademie-Verlag, DDR – 108 Berlin, Leipziger Straße 3–4, Fernruf 2 23 62 21 oder 22 36 · Telex Nr. 11 44 20, Postscheckkonto: Berlin 350 21 · Die Zeitschrift für Meteorologie erscheint jährlich in einem Band mit 6 Heften · *Herstellung:* IV/2/14 VEB Druckerei »Gottfried Wilhelm Leibniz«, 445 Gräfenhainichen/DDR · Veröffentlicht unter der Lizenznummer 1281 des Presseamtes beim Vorsitzenden des Ministerrates der Deutschen Demokratischen Republik · Nachdrucke, Vervielfältigungen, Verbreitungen und Übersetzungen in fremde Sprachen des Inhalts dieser Zeitschrift – auch auszugsweise mit Quellenangabe – bedürfen der schriftlichen Genehmigung des Verlages. Printed in the German Democratic Republic.

Namen- und Sachregister der Zeitschrift für Meteorologie, Band 27, 1977

A

B

C

D

L

M

S

T

U

Zeitschrift für Meteorologie

Heft 1 · 1978 · Band 28

Im Auftrage der
Meteorologischen Gesellschaft
der Deutschen
Demokratischen Republik

herausgegeben von
K. Bernhardt, Berlin,
E. A. Lauter, Berlin und
H. Pleiß, Dresden

Akademie-Verlag · Berlin

Zf Meteor. Bd. 28 Heft 1 S. 1–64 · Berlin 1978
EVP 8,– M
34 126

Bezugsmöglichkeiten der Zeitschrift für Meteorologie

Bestellungen sind zu richten

– *in der DDR* an den Postzeitungsvertrieb, an eine Buchhandlung oder an den Akademie-Verlag, DDR – 108 Berlin, Leipziger Straße 3–4

– *im sozialistischen Ausland* an eine Buchhandlung für fremdsprachige Literatur oder an den zuständigen Postzeitungsvertrieb

– *in der BRD und Westberlin* an eine Buchhandlung oder an die Auslieferungsstelle KUNST UND WISSEN, Erich Bieber, 7 Stuttgart 1, Wilhelmstraße 4–6

– *in Österreich* an den Globus-Buchvertrieb, 1201 Wien, Höchstädtplatz 3

– *im übrigen Ausland* an den Internationalen Buch- und Zeitschriftenhandel; den Buchexport, Volkseigener Außenhandelsbetrieb der Deutschen Demokratischen Republik, DDR – 701 Leipzig, Postfach 160, oder an den Akademie-Verlag, DDR – 108 Berlin, Leipziger Str. 3–4.

Zeitschrift für Meteorologie

Herausgeber: Im Auftrage der Meteorologischen Gesellschaft der Deutschen Demokratischen Republik herausgegeben von Prof. Dr. Karlheinz Bernhardt, Berlin; Prof. Dr. Ernst August Lauter, Berlin und Prof. Dr. Hermann Pleiß, Dresden.

Verlag: Akademie-Verlag, DDR – 108 Berlin, Leipziger Str. 3–4; Fernruf 2 23 62 21 oder 2 23 62 29
Telex-Nr.: 11 44 20; Postscheckkonto: Berlin 35021; Bank: Staatsbank der DDR, Berlin, Kto.-Nr.: 6836-26-20712.

Chefredakteur: Dozent Dr. Rolf Haake, Potsdam

Anschrift der Redaktion: Zeitschrift für Meteorologie, DDR – 15 Potsdam, Kleine Weinmeisterstr. 14; Fernruf: Potsdam 2 54 18.

Veröffentlicht unter der Lizenznummer 1281 des Presseamtes beim Vorsitzenden des Ministerrates der Deutschen Demokratischen Republik.

Gesamtherstellung: VEB Druckerei »Gottfried Wilhelm Leibniz«, DDR – 445 Gräfenhainichen.

Erscheinungsweise: Die Zeitschrift für Meteorologie erscheint jährlich in einem Band mit 6 Heften.
Bezugspreis je Band 90,– M zuzüglich Versandspesen (Preis für die DDR 48,– M);
Preis je Heft 15,– M (Preis für die DDR 8,– M).

Bestellnummer dieses Heftes: 1013/28/1

AN (EDV) 7725

Berichtigung zum Beitrag

J. Kluge, K. H. Stahnke-Jungheim und Ch. Vollbrecht

Die Verwendung von Winddaten bei der operativen numerischen Analyse des Bodendruckfeldes (Z. Meteor. 27 (1977), H. 3, 147–155

(33) $$K_p = \overline{p'_k p'_i}^N = (1 + r_{ki}/L)\, e^{-r_{ki}/L}$$

(37) $$K_{vv} = \overline{\frac{\partial p'_i}{\partial x} \frac{\partial p'_j}{\partial x}}^N = \left(1/L^2 - \frac{(x_i - x_j)^2}{r_{ij} L^3}\right) e^{-(r_{ij}/L)} + \delta_{ij} \zeta_v$$

(38) $$K_{uu} = \overline{\frac{\partial p'_i}{\partial y} \frac{\partial p'_j}{\partial y}}^N = \left(1/L^2 - \frac{(y_i - y_j)^2}{r_{ij} L^3}\right) e^{-(r_{ij}/L)} + \delta_{ij} \zeta_u$$

Korrektur

In der Arbeit Haase/Leidreiter in Heft 5, Band 27 muß es auf Seite 281, Zeile 6 richtig heißen „unterdurchschnittlichen Werten"

1013/28/1

ISSN 0084-5361

Zeitschrift für Meteorologie

Heft 1 · 1978 · Band 28

Nachruf

551.5:92

S. P. Chromow †
geb. 4. 9. 1904 gest. 29. 4. 1977

Am 29. April 1977 ist Professor Dr. Ssergej Petrowitsch Chromow in Moskau im 73. Lebensjahr nach einer kurzen Krankheit gestorben. Der Verstorbene war ein bekannter Wissenschaftler auf dem Gebiet der synoptischen Meteorologie und der Klimatologie.

Seine wissenschaftliche Laufbahn begann im Jahre 1922 als er sein Hochschulstudium an der Physikalisch-mathematischen Fakultät der Staatlichen Universität in Moskau begonnen hatte. Im Jahre 1928 promovierte er als Geophysiker. Noch als Student arbeitete er seit dem Jahre 1924 im Aerologischen Observatorium des Staatlichen Geophysikalischen Instituts und trat zwei Jahre später zum Moskauer Bezirks-Wetterdienst über, wo sich sein wissenschaftliches Hauptinteresse in der synoptischen Meteorologie und im praktischen Wetterdienst herausbildete.

Zu dieser Zeit einer stürmischen Entwicklung der von V. Bjerknes eingeführten Methoden der frontologischen Wetterkarten-Analyse begann Chromow diese Arbeitsmethoden im praktischen Wetterdienst einzuführen. Bei der Errichtung des Zentralen Wetterdienstes der UdSSR in Moskau wurde Chromow dort Leiter der Forschungsabteilung, die er selbst gründete.

In den dreißiger Jahren wurde die Bjerknessche frontologische Analyse der Wetterkarten in der Sowjetunion eingeführt, wobei Chromow in der Arbeitsgemeinschaft junger Synoptiker eine leitende Rolle spielte. Er nahm teil an den damals in Moskau organisierten Vorlesungen von T. Bergeron, welcher bei dieser Gelegenheit zweimal längere Zeit in der Sowjetunion weilte. Unter der Redaktion von Chromow wurde damals in russischer Sprache der zweite Teil von Bergerons Arbeit „Dreidimensional verknüpfte Wetteranalyse" und „Vorlesungen über Wolken in der praktischen Analyse der Wetterkarte" veröffentlicht. Seit dem Jahr 1932 hielt Chromow oft Vorlesungen über die neue Methode der Wetteranalyse zwecks weiterer Ausbildung von Synoptikern in der Praxis. In diesen Zeitraum fällt auch der Beginn der systematischen pädagogischen Arbeit von Chromow. Seine wichtigste Arbeit war die „Einführung in die synoptische Wetteranalyse", die im Jahre 1934 erschien. Die zweite, wesentlich erweiterte Ausgabe des Buches stammte aus dem Jahre 1937; nahezu gleichzeitig damit wurde eine tschechische Übersetzung des Werkes in Prag herausgegeben. Im Jahre 1940 wurde eine deutsche Bearbeitung im Springer-Verlag in Wien veröffentlicht.

Im zweiten Weltkrieg beteiligte sich Chromow an der Ausbildung von Meteorologen für die Arbeit unter Kriegsverhältnissen. Gegen Ende des Jahres 1944 übersiedelte Chromow nach Leningrad, wo er zum Professor und Leiter des Lehrstuhls des Hydrometeorologischen Instituts bestellt wurde; seit 1947 arbeitete er auch an der Geographischen Fakultät der Leningrader Universität als Leiter des Lehrstuhls für Meteorologie und Klimatologie

Im Jahre 1948 erschien ein umfangreiches Werk von Chromow unter dem Titel „Grundzüge der synoptischen Meteorologie". Diese Arbeit wurde von der Geographischen Gesellschaft der Sowjetunion durch die Verleihung einer Goldmedaille ausgezeichnet.

Im Jahre 1953 kehrte Chromow nach Moskau zurück und betätigte sich dort bis zu seinem Tod als Mitglied und längere Zeit als Leiter des Lehrstuhls für Meteorologie und Klimatologie.

In den Jahren 1956–57 nahm Chromow an der Sowjetischen Expedition in der Antarktis auf dem Schiff „Ob" teil. Seine ausgedehnte pädagogische und wissenschaftliche Arbeit vereinigte Chromow erfolgreich mit öffentlicher Tätigkeit. Er war Mitglied vieler wichtiger wissenschaftlicher Organisationen und Redaktionen von Fachzeitschriften.
Im Jahre 1956 veröffentlichte Chromow gemeinsam mit L. I. Mamontowa das „Meteorologische Wörterbuch", das 7 Jahre später in einer Neubearbeitung erschien; nunmehr wird eine dritte Ausgabe dieses Werkes vorbereitet.

Als Anerkennung und Würdigung seiner vielfältigen Tätigkeit im wissenschaftlichen, organisatorischen und pädagogischen Bereich wurden Professor Chromow viele Ehrungen zuteil. Er wurde auch zum Ehrenmitglied der Meteorologischen Gesellschaften der Deutschen Demokratischen Republik, der Tschechoslowakischen Sozialistischen Republik und der Ungarischen Volksrepublik gewählt.

N. Končck

551.509.313:681.3
551.507.362.2

Spatial Analysis of Temperature and Geopotential Fields on the Basis of data from remote sounding of the Atmosphere

By O. M. Prokovskij and E. E. Ivanykin

Zusammenfassung: Das Verfahren zur räumlichen Assimilation indirekter meteorologischer Informationen [6–8] wird auf das Problem der Analyse des Temperatur- und Geopotentialfeldes aus Satelliten-Meßdaten der thermischen Ausstrahlung angewandt. Es werden mehrere Assimilations-Schemata diskutiert: räumliche Extrapolation und Interpolation, optimale Filterung eines Datensatzes von Satellitenmessungen, räumliche Anpassung aerologischer Informationen und Daten der indirekten Sondierung bei der objektiven statistischen Analyse des Geopotentialfeldes. Die Ergebnisse der numerischen Modellierung der Assimilations-Schemata auf Elektronenrechenmaschinen werden mitgeteilt. Es werden sowohl die Abhängigkeit der Genauigkeit der erhaltenen Abschätzungen von der Dimension des räumlichen Filters als auch die Stabilität der angewandten Algorithmen gegenüber Fehlern der Eingangsinformation untersucht. Der vorgeschlagene Ansatz ermöglicht, verglichen mit konventionellen Interpretationsverfahren, eine optimalere Nutzung von Satelliteninformationen.

Summary: The technique of spatial assimilation of indirect meteorological information [6–8] is applied to the analysis of the temperature and geopotential field on the basis of satellite-borne measurements of outgoing thermal radiation. A number of assimilation schemes is discussed: spatial extrapolation and interpolation, optimal filtration of a set of data of satellite measurements, spatial adjustment of aerological information and remote sounding data in the objective statistical analysis of the geopotential field. The results of numerical modelling of the assimilation schemes by electronic computers are given. The dependence of the accuracy of estimates obtained on the dimension of the spatial filter is investigated as well as the stability of the applied algorithms as compared with the errors in the initial information. It is shown that the proposed approach permits better use to the made (compared with the conventional interpretation techniques) of satellite information.

The techniques for the interpretation of the characteristics of the fields of outgoing thermal radiation of the earth-atmosphere system designed to obtain the vertical temperature distributions in the atmosphere imply, first of all, meteorological applications. In using conventional methods of the inversion of remote sounding data, the obtained information has only a limited value for numerical forecasting purposes [5], which is due to a relatively low accuracy of temperature determination characteristic of modern interpretation methods. Procedures of objective analysis [3] designed for the processing of aerological information are not optimum when applied to the results of the interpretation of satellite measurements whose errors cannot be considered to the spatially independent [2].

Conventional interpretation methods imply an independent processing of the data of satellite measurements referring to the different points of the meteorological field. Correlation radii of random fields of the main meteorological elements (e.g. temperature and geopotential) are of the order of 1500–2000 km, a dozen times the value of the distance between the reference points of neighbouring satellite measurements. The restriction of the statement to conventional interpretation methods leads to their non-optimality (from the statistical point of view.

The linearity of the inverse problem considered (with respect to the parameters sought) creates the basis for the generalization of interpretation techniques. Based upon the general methodology involving the statistical analysis of a spatial set of the data of remote sounding, their optimum adjustment to the aerological information and the interpolation in the nodes of the regular three-dimensional grid [6, 7], numerical experiments on electronic computers were made in connection with the retrieval of the temperature field for a number of schemes of spatial assimilation for certain meteorological situations. The results of the investigation show that the use of the information on close spatial correlations of the temperature field appears to be an important factor in increasing the accuracy of the estimates obtained on the basis of satellite data.

One should bear in mind that the numerical forecasting models often make use not of the temperature distribution but of the geopotential profile which is a linear function of temperature. Moreover, in many cases the information on spatial gradients of the geopotential field is indispensable. Therefore, in view of the prospects of using the data of remote sounding as an initial meteorological information it is important to study the questions concerning the estimation of the accuracy of the determination of the geopotential for standard isobaric surfaces with the aid of different assimilation schemes. The present paper is devoted to the consideration of the main elements of the objective analysis of the temperature and geopotential field using the data of satellite measurements with and without aerological information.

1. The problem

The problem of assimilation of meteorological information, its interpretation and analysis is closly connected with the filtration and estimation problems. Let us formulate the filtration problem in its conventional form.

Consider $\{x(t), t \in T\}$ and $\{n(t), t \in T\}$ as two random processes, which we shall call a signal and noise, respectively. Let us assume that the sum $y(t) = x(t) + n(t)$ is the result of the measurement of the signal $x(t)$ in thetime interval $[t_0, t_1]$. Based upon the given $y(t)$ $(t_0 \leqq t \leqq t_1)$ it is necessary to obtain the best estimate of the value of the signal at the time $t=\bar{t}$. Let us denote by Y a set of observational data. In the discrete case Y is a vector. For processes continuous in time, Y is a function representing the realization of the random process $y(t)$. Let X be a set of values of the signal $x(t)$. The filter is a function expressing Y by X. The value of this function at the concrete Y is called an estimate and is denoted by $\hat{x}$.

For a full description of the problem it is necessary to determine „signal" and „noise", i.e. the criterion from which the best estimate and the class of allowed filters are determined. The signal and the noise may be determined in serveral ways. They can be represented as random processes or as vectors, and can be described with the aid of covariance functions, spectral densities or stochastic equations. The quality of the estimates is usually characterized by the loss function. The loss function is determined as the real function $L(x)$ with the following properties: 1) $L(x)\geqq 0$ 2) $L(x)=L(-x)$ 3) $L(x)$ does not decrease with increasing $||x||$. Then the losses become the random value $L(x-\hat{x})$. The best estimate $\hat{x}$ minimizes the mean losses $E\{L(x-\hat{x})\}$ (E is the symbol denoting mathematical expectation).

Let us dwell upon the theoretical principles of obtaining the best values. All the necessary statistical information on the value $x(\bar{t})$ extracted from the observations $y(t)$ is contained in the conditional distribution of probabilities

$$F(\tilde{x}/\tilde{y})=P\{x(\bar{t})\leqq\tilde{x} \mid y(t)=\tilde{y}(t),\ t_0\leqq t\leqq t_1\}$$

We shall denote the density of this distribution by $f(\tilde{x}/\tilde{y})$. To find the best filter, it is necessary to find the function $\hat{x}=\hat{x}(\tilde{y})$ minimizing $E\{L(x-\tilde{x})\}$. Using the well-known properties of the operation of conditional mathematical expectation we obtain

(1) $$E\{L(x-\tilde{x})\}=E\tilde{y}[E\{L(x-\hat{x})/\tilde{y}\}]\ .$$

Here $E\{\ldots/\tilde{y}\}$ denotes the conditional mathematical expectation at $y(t)=\tilde{y}(t)$, $t_0\leqq t\leqq t_1$. Therefore, the minimization (1) over all filters $\hat{x}=\hat{x}(\tilde{y})$ is equivalent to the determination of the minimum of the conditional mean losses

(2) $$E\ \{L(x-\tilde{x})/\tilde{y}\}=\int_{-\infty}^{\infty} L(\tilde{x}-\hat{x})\ f(\tilde{x}/\tilde{y})\ d\tilde{x}\ .$$

The rule of obtaining the best estimate in the sense of (2) proceeds from the fundamental result [9] to be formulated as follows:

A. Let us assume that the loss function $L(x)$ satisfies the above properties 1)–3). Let also the density function $f(\tilde{x}/\tilde{y})$ be symmetric with respect to the conditional mathematical expectation $m=\int_{-\infty}^{\infty}\tilde{x}\ f(\tilde{x}/\tilde{y})\ d\tilde{x}$. Then the estimate minimizing the mean losses (2) is fixed by the conditional mathematical expectation:

(3) $$\hat{x}=\hat{x}(\tilde{y})=E\ \{x/\tilde{y}\}=\int_{-\infty}^{\infty} x\ f(x/\tilde{y})\ dx\ .$$

We shall point out to some consequences following from this result.

From (3) it follows that the particular choice of the loss function is not essential. The conditional mean $\hat{x}$ is the best estimate with respect to any admissible function $L(x)$ (i.e., satisfying conditions 1)–3)). Most frequently used is the quadratic loss function $L(x)=||x||^2$ which is treated below. If the processes x and n are Gaussian, the conditional mathematical expectation $E\ \{x/\tilde{y}\}$ is the linear function $\tilde{y}$. Thus a linear filter is the optimum one relative to the Gaussian random functions.

In meteorological applications, there is the important case when the signal and the noise are random processes of the second order. This means that we have information only on the first two moments of the data of random function. Then the random functions considered are equivalent to Gaussian processes with the same first and second moments. Thus, in this case the optimum filter is linear and coincides with what could be obtained if the processes of the second order were substituted by the corresponding Gaussian processes. If we do not deal with function but with random Gaussian vectors, the expressions for the optimum filters appear sufficiently simple. Let us formulate the most important results [1].

B. Let x and y be the vectors having a joint Gaussian distribution with the mathematical expectation $\begin{bmatrix}E(x)\\E(y)\end{bmatrix}$ and the covariance matrix $\begin{bmatrix}R_x & R_{xy}\\R_{yx} & R_y\end{bmatrix}$. Then the conditional distribution $P(x/y)$ is characterized by the mathematical expectation:

(4) $$E(x/y)=E(x)+R_{xy}\cdot R_y^{-1}\ [y-E(y)]$$

and the covariance matrix

(5) $$R_{x/y}=R_x-R_{xy}\cdot R_y^{-1}\cdot R_{yx}$$

$[x-E(x)]$ and y are statistically independent. Let us point out to an important partial case of this statement.

C. Let x, y, z be random Gaussian vectors. y and z are statistically independent. Then the conditional distribution $P(x/y,z)$ is characterized by the mathematical expectation

(6) $$E(x/y,z)=E(x/y)+E(x/z)-E(x)$$

and the covariance matrix

(7) $$R_{x/y,z}=R_x-R_{xy}\cdot R_y^{-1}\cdot R_{yx}-R_{xz}\cdot R_z^{-1}\cdot R_{zx}\ .$$

Let us interpret the given results from the viewpoint of the estimation theory. Consider the problem of the estimation of x from the observations y so that the quadratic loss function reaches its minimum, i.e., the root-mean square error of the best estimate is least. According to statement B, the best estimate $\hat{x}$ which is linear is determined by equation (4). The estimation error is characterized by the covariance matrix (5). Note, that if the vectors x and y are not Gaussian, the linear filter (4) provides the estimate which is the best in the class of linear estimates (i.e., minimizing the quadratic criterion $E(||x-\hat{x}||^2)$).

In the case of independent measurements y and z the estimation of the wanted parameter x is made according to (6) based upon two independent filters. The covariance matrix (7) of the estimate $\hat{x}$ is characterized by subtraction of two independent terms.

Such are the general principles the modern filter methods are based upon. Further we shall consider the possibilities of their realization in the algorithms of space-time assimilation of meteorological information. Let us dwell upon the analysis of two main types of information appropriate for models of numerical weather forecasting.

The aerological information obtained at a network of meteorological stations is presently the basis of objective analysis of fields of the main meteorological elements [3]. Prospects of the use of global information from meteorological satellites require the development of corresponding methods for the analysis of input data. The problem of the agreement of both types of meteorological information is determined by their heterogeneity. Aerological information is direct. The radiosounding technique permits the distribution of meteorological parameters to be obtained immediately. Data of remote sounding of the atmosphere from satellites yield indirect information about them. The measured values are only functionals of the wanted distributions. The aerological nétwork is distributed very non-uniformly over the earth's surface. Measurements are carried out within considerable time intervals. As a rule, they are performed twice a day (in standard terms: 00.00 Z and 12.00 Z).

Satellite information uniformly covers the whole surface of the globe. It is virtually continuous in time. Therefore, when speaking about objective analysis of both kinds of meteorological data one may consider two problems. The first is spatial analysis of aerological data and their optimum adjustment to satellite information referring to the minor temporal vicinities of standard terms of radiosounding. The second problem is related to space-time assimilation of the data of satellite measurements. In the following sections, both the formulated problems will be considered.

2. Objective analysis of the data of direct and indirect measurements

If the input information is the data of direct (radiosounding) measurements of meteorological elements the technique of the objective statistical analysis developed by *L. S. Gandin* [3] permits the obtained information to be interpolated and extrapolated at an optimum in the nodes of the regular three-dimensional grid, for immediate use in numerical forecasting models. The given method is based upon the results of linear filters for random values of the second order, as treated in the preceding section. The filtration principles of optimum filtering may also be used for the analysis of satellite data. The statistical regularization method widely used in the practice of interpretation of remote sensing data, designed for the processing of individual spatial measurements from satellites, constitutes the optimum linear filter. Therefore, the statistical analysis of the meteorological field over the arbitrary finite set of data of direct and indirect measurements implies the presence of statistical regularization and optimum extrapolation elements in the corresponding optimum filter. The results obtained are a generalization of both the above-mentioned methods.

The numerical solution of the problem of indirect thermal sounding of the atmosphere is connected with the consideration of a system of linear algebraic equations

$$\Delta I(x) = A(x) \cdot \Delta T(x) + \varepsilon(x) \tag{8}$$

which relates variations of the vector of measured characteristics of outgoing radiation $\Delta I(x)$ (x – is a spatial coordinate) to the corresponding fluctuations of the vector of values of the vertical temperature distribution ΔT. A is the matrix approximating the integral operator of transfer equation, $\varepsilon(x)$ is the random value characterizing both stochastic errors in the used model and errors of measurements. Let, for certainty, the vectors ΔI and ΔT have the dimensions m and n, respectively.

The geopotential of the isobaric surface with the pressure $p = \bar{p}$ is determined by

$$H(\bar{p}) = H(p_0) + c \cdot \int_{\ln \bar{p}}^{\ln p_0} T_v d \ln p \tag{9}$$

where $c = R/g$ (R is the specific gaseous constant of the air, g is the gravity acceleration), T_v is the virtual temperature represented by the approximate equation $T_v\ (p) = T(p) + g(p)$. ($g(p)$ is the mixing ratio of water vapor (g/kg) in the atmosphere at the pressure level P) [10]. Assuming P_0 to be equal to the value of surface pressure P_s, it is possible to consider that $H(p_s) = 0$. Using (9) we shall write the total variation for the value $H(\bar{p})$ in the following way.

$$\Delta H(\bar{p}) = c \cdot \left[\int_{\ln \bar{p}}^{\ln p_s} \Delta T(p) d \ln p + T_v(p_s) \cdot \Delta \ln p_s \right] \tag{10}$$

with ΔT_v the subscript v is dropped since humidity variations whose contribution is much less than that of the dominating temperature fluctuations may be neglected. The remote sounding method does not permit to infer the variations of the surface pressure characterizing the second term in (10). The possibilities of the method following from (8) are confined to monitoring the variations of the vertical temperature profile. The data on $\Delta \ln p$ must be obtained independently. For example, objective analysis of the surface pressure field from data of direct measurements may serve as a basis for obtaining similar information. However, since this is not connected with the subject of our investigation (analysis of satellite information) we shall restrict ourselves to the estimation of geopotential variations from fluctuations of the temperature distribution by which they are controlled.

Similar to (8) we shall consider ΔT to be the n-dimensional vector $\Delta T=(\Delta T(p_1),\ldots,\Delta T(p_n))$, let $\Delta T(p_1)=$ $=\Delta T(p_s)$. In this case the $(n-1)$-dimensional vector $\Delta H=(\Delta H(p_2),\ldots,\Delta H(p_n))$ is connected with the ΔT matrix relation

(11) $\quad \Delta H=B\cdot\Delta T$

which contains a triangular matrix B of the dimension $(n-1)\times n$, $B=\{b_{ij}\}$ $(i=2,\ldots n,\ j=1,\ldots,n)$, $b_{ij}=c\cdot c_j^i$ for $j\leqq i$ and $b_{ij}=0$ for $j\geqq i$. The values c_j^i $(j=1,\ldots,i)$ constitute the coefficients of the quadratic formula approximating the integral in (9) at $\bar{p}=p_i$ $(i=2,\ldots,n)$.

Let us determine the problem of statistical analysis of the results of measurements $\Delta I(x)$ made at points with coordinates $x_1,\ldots,x_N$. Our aim is to obtain the best statistical estimates for the deviations of meteorological elements from their mean values $\Delta T(x_0)=T(x_0)-\bar{T}(x_0)$ and $\Delta H(x_0)=H(x_0)-\bar{H}(x_0)$ at an arbitrary point of the field with coordinates x_0. Note, that if $A(x_i)=I$ (I is the matrix of identical transformation) the stochastic model (8) describes the results of aerological measurements. We assume that all matrix operators $A(x_i)$, the vectors of mean values $\bar{T}(x_i)$ and the covariance matrices $R_T(x_i,x_j)$, $R_T(x_i)$, $R_\varepsilon(x_i,x_j)$ $(i,j=0,1,\ldots,N)$ are determined.

We combine the results of measurements $\Delta I(x_i)$ in the single block vector

$$\Delta I(x_1,\ldots,x_N)=\begin{vmatrix}\Delta I(x_1)\\ \vdots\\ \Delta I(x_N)\end{vmatrix}.$$

In order to obtain the necessary estimate we make use of the apparatus of multivariate statistical analysis. According to (4,5), for the best statistical estimate minimizing quadratic deviations the expression

(12) $\quad \Delta\hat{T}(x_0)=K_T(x_0\mid x_1,\ldots,x_N)\cdot\Delta I(x_1,\ldots,x_N),$

containing the prediction operator

(13) $\quad K_T(x_0\mid x_1,\ldots,x_N)=R_{TI}\cdot R_I^{-1}$

is valid.

Covariance matrices on the righthand side of (13) have a block structure. The quadratic matrix R_I contains $N\times N$ blocks in the form of $\{R_I\}_{ij}=A(x_i)\cdot R_T(x_i,x_j)\cdot A^*(x_j)+R_\varepsilon(x_i,x_j,)$ $(i,j=1,\ldots,N)$.

The cross covariance matrix R_{TI} of the estimated vector $\Delta T(x)$ and the input data constitute a row consisting of N blocks in the form of $R_T(x_0,x_i)\cdot A^*(x_i)$ $(i=1,\ldots,N)$.

The quality of the estimate (12) is characterized by the matrix of covariances of the random vector $\Delta\hat{T}(x_0)$

(14) $\quad \hat{R}_T(x_0)=R_T(x_0)-K_T(x_0\mid x_1,\ldots,x_N)\cdot R_{TI}^*.$

Since the variations ΔT and ΔH are interconnected by a linear relation, the first two moments of the random vector $H(x)$ are described by

(15) $\quad \bar{H}(x)=B\cdot\bar{T}(x),\ R_H(x,x')=B\cdot R_T(x,x')\cdot B^*.$

Similarly, taking account of (11) we arrive at the estimates for the geopotential at the point:

(16) $\quad \Delta\hat{H}(x_0)=B\cdot K_T(x_0\mid x_1,\ldots,x_N)\cdot\Delta I(x_1,\ldots,x_N)$

(17) $\quad \hat{R}_H(x_0)=R_H(x)-B\cdot K_T(x_0\mid x_1,\ldots,x_N)\cdot R_{TI}^*\cdot B^*.$

The predictor $K_T(x_0\mid x_1,\ldots,x_N)$ using the data of N independent measurements is called the filter of the order N. Let us point out to particular cases of the descibed technique of statistical analysis. When the information is fixed at one point $(N=1)$, $x_0\neq x_1$ the formulated procedure constitutes the optimum extrapolation. At $N>1$ the algorithm is reduced to optimum interpolation. If the meteorological elements $T(x)$ and $H(x)$ are fixed on more than one isobaric surface, the assimilation mechanism performs the optimum adjustment of the values of the corresponding meteorological fields [3]. When the supporting points $x_1,\ldots,x_N$ are chosen by turns as x_0 the algorithm considered is reduced to optimum spatial adjustment (filtering) and provides the best estimates of meteorological elements at each of these points based upon the entire set of measurements $\Delta I(x_1),\ldots,\Delta I(x_N)$. Note, that one of the conventional interpretation algorithms (statistical regularization method) constitutes of a first order filter. Assuming $A(x_i)=I$ we arrive at the method of objective statistical analysis developed by *L. S. Gandin* [3]. If the blocks of the vector ΔI contain both aerological data and results of satellite measurements the method considered permits optimum adjustment of both kinds of meteorological information.

A serious difficulty in the course of numerical solution of the assimilation problem on electronic computers with the aid of the above formal procedure is the inversion of the block matrix, due to its large dimension. In practice, while analysing the meteorological field referring even to a comparatively small area of the earth's surface, it may be necessary to use a considerable amount of measured data. In order to overcome this difficulty the use of the recursive algorithm of successive inversion R_I over blocks is proposed. It is based upon the matrix identity easily checked:

(18) $$\begin{Vmatrix}A & C\\ C^* & B\end{Vmatrix}^{-1}=\begin{Vmatrix}\tilde{A} & \tilde{C}\\ \tilde{C}^* & \tilde{B}\end{Vmatrix};\ \tilde{A}=(A-C\cdot B^{-1}\cdot C^*)^{-1},\tilde{C}=-\tilde{A}\cdot C\cdot B^{-1},\ \tilde{B}=B^{-1}+B^{-1}\cdot C^*\cdot\tilde{A}\cdot C\cdot B^{-1}.$$

Let us apply (18) for inversion R_I choosing as B the last diagonal block R_I and as C and C^* the last block column and the row R_I without the last block. In place of the block $\tilde{B}$ we shall obtain the matrix expression containing the matrix inverse to the matrix $A(x_N) \cdot R_T^{-1} \cdot A^*(x_N) + R_\varepsilon$. The inversion of this matrix due to the presence of the regularization term $R_\varepsilon(x_N)$ is quite stable. This is confirmed by the practice of the solution of inverse problems by the statistical regularization method. The computation of the block $\tilde{A}$ requires the computation of the block matrix of dimension $(N-1)\times(N-1)$ with the following block structure: $\{\tilde{A}\}_{ij} = A(x_i) \cdot [R_T\ (x_i, x_j) - R_T(x_i, x_N) \cdot (A(x_N) \cdot R_T \cdot A^*(x_N) + R_\varepsilon)^{-1} \cdot A^*(x_N) \cdot R_T(x_N, x_i)] \cdot A^* + R_\varepsilon \cdot \delta_{ij}$ $(i,j = 1, \ldots, N-1$; δ_{ij} is the Kronecker symbol). The procedure mentioned should be applied to this matrix again and after the performance of the inversion of the last diagonal block $\tilde{A}_{N-1,N-1}$ which also contains the regularization term $R_\varepsilon(x_{N-1})$ we arrive at the problem of the computation of the inverse block matrix $(N-2)\times(N-2)$. After the $(N-1)$-step of the recursive algorithm we have to invert the matrix the dimension of which is equal to the dimension of the first diagonal block R_I. Thus, to find the inverse matrix for R_I it is necessary to invert N of the regularization matrix operators whose dimension is equal to that of the diagonal blocks R_I. Since the matrix R_I does not depend upon the current point it must be inverted only once. The matrix R_{TI} taking account of the dependence on x_0 is computed without any difficulties.

Thus, the procedure of the assimilation of the data of N measurements is equivalent in its numerical relation to N individual inversions of the data $\Delta I(x_i)$ $(i = 1, \ldots, N)$ for obtaining $\Delta\hat{T}(x_0)$. However, the application of the assimilation algorithm permits to obtain the estimate at an arbitrary point of the fields T, H including the nodes of the regular three-dimensional grid for further use in the prediction schemes.

3. Optimum statistical procedures of time-space assimilation of indirect meteorological information

The increasing role of asynoptical information which in contrast to synoptical information is practically continuous in time requires the consideration of temporal assimilation of this kind of data.

In the numerical solution of the inverse problem the initial integral equation is approximated by the system of linear algebraic equations. Here we shall write it in the following form:

(19) $$A(t) \cdot \Delta Z(t) = \Delta y(t) + \varepsilon(t)$$

which relates the variations of the vector of measured radiation characteristics $\Delta y(t)$ at the time t to the corresponding variations of the vector of values of the wanted meteorological element $\Delta Z(t)$. Here $\varepsilon(t)$ is the random error of measurements. For certainty, we shall consider that A is the rectangular matrix $m\times n$ and that the dimensions of the vectors ΔZ and Δy are equal to n and m, respectively.

Now we assume that the projection of the trajectory of scanning of the measuring device on the earth's surface is described by $x = x(t)$. Here x is the vector of the coordinates of the current point to which the data of measurements $y(t)$ refer. In this case it is possible to determine the random function $Z(t) = Z(x(t), t)$ constituting the „trajectory" in the random space-time field of the wanted meteorological elements Z. The stochastic structure of the random function $Z(t)$ is determined uniquely by the probability parameters of the initial meteorological field. Its first and second moments can be found from the equations

(20) $$\bar{Z}(t) = \bar{Z}(x(t), t); \quad R_Z(t_1, t_2) = R_Z[x(t_1), t_1; x(t_2), t_2].$$

Note, that the function of mean values $\bar{Z}(t)$ and the covariance matrix function $R_z(t_1, t_2)$ fully characterize the space-time structure of the meteorological field along the trajectory of scanning. We shall assume that the vector $\varepsilon(t)$ is not correlated with the components $Z(t)$ and has zero mean values. We shall also assume that the corresponding covariance function is described by the equation $R_\varepsilon(t_1, t_2) = R_\varepsilon(t_1) \cdot \delta(t_1 - t_2)$. The first two moments of the random field of signals Y are determined from

(21) $$\bar{y}(t) = A(t) \cdot Z(t); \quad R_Y(t_1, t_2) = A(t_1) \cdot R_Z(t_1, t_2) \cdot A^*(t_2) + R_\varepsilon(t_1, t_2)$$

(* – is the sign of matrix conjugation).

The determinations introduced permit the formulation of the main problem in its most general form. We consider that indirect measurements are performed continuously in the time interval $[t_0, t_1]$. It is necessary to obtain the best statistical estimate at the arbitrary current point S based upon a set of data on the realization of the random function $y(t)$ $(t_0 \leq t \leq t_1)$ using linear operations. Since in meteorological practice only two first moments of random fields are considered, the procedure of linear estimation appears optimum. The random field determined by two moments is equivalent to the Gaussian field for which the optimal predictors are in the class of linear functionals.

Assuming that the functions $Z(t)$ and $y(t)$ are centered by their means, i.e., $\bar{Z}(t) = \bar{y}(t) = 0$ we shall seek the solution of the problem in the form of the following linear estimate:

(22) $$\hat{z}(s) = \int_{t_0}^{t_1} K(s, t) \cdot y(t)dt$$

The integral form of the vector or matrix expression is the vector or matrix of integrals of the corresponding components. The function $K(s, t)$ still undetermined can be found from the optimization condition of a certain criterion related to the vector of differences $\delta\hat{Z}(s) = Z(s) - \hat{Z}(s)$. Let us choose as a criterion the most widely used measure of the variation of random vector values, viz., the generalized dispersion $\delta\hat{Z}(s)$. In this case after simple calculations we have the following expression for the covariance matrix of scattering the value (22):

(23) $$D(\delta\hat{Z}(s)) = E(\delta\hat{Z}(s) \cdot \delta\hat{Z}(s)^*) = \int_{t_0}^{t_1}\int_{t_0}^{t_1} K(s, t)\{A(t) \cdot R_Z(t, \tau) \cdot A^*(\tau) + R_\varepsilon(t, \tau)\} \cdot K^*(s, \tau)\, dt\, d\tau + R_Z(s, s) - \int_{t_0}^{t_1} K(s, t) \cdot A(t) \cdot R_Z(t, s)dt - \int_{t_0}^{t_1} R_Z(s, t) \cdot A^*(t) \cdot K^*(s, t)dt$$

In obtaining (23) the properties of the linearity of the operations of integration and mathematical expectation are used.

After fixing the value $t=s$ we arrive at the classical problem of calculus of variations for the matrix function $K(s, t)$ of one variable t in the interval $[t_0, t_1]$. It is necessary to find the function K realizing the minimum of the functional of the conventional form

(24) $$\int_{t_0}^{t_1} F(t, K(s, t))\,dt$$

where

(25) $$F(t, K(s, t)) = \int_{t_0}^{t_1} K(s, t)\{A(t)\cdot R_Z(t, \tau)\cdot A^*(\tau) + R_\varepsilon(t, \tau)\}\times K^*(s, \tau)\,dt - -(K(s, t)\cdot A(t)\cdot R_Z(t, s) + R_Z(s, t)\cdot A^*(t)\cdot K^*(s, t)).$$

Taking account of (25), after some transformations we obtain for (24) Euler's equation in the following form:

(26) $$\int_{t_0}^{t_1} K(s, t)\{A(t)\cdot R_Z(t, \tau)\cdot A^*(\tau) + R_\varepsilon(t, \tau)\}\,dt = R_Z(s, \tau)\cdot A^*(\tau).$$

Formula (26) describes a system of Fredholm integral equations of the 1-st kind relative to the matrix function $K(s, t)$ sought at each s. System (26) must be satisfied by the function $K(s, t)$ at all t from the interval $[t_0, t_1]$ Substituting the obtained solutions in (23) we find the covariance matrix of the best estimate.

Equations in the form of (26) usually emerge in filtration problems and allow an analytical solution only in the stationary case when the methods of Fourier analysis can be applied. In the general case, effective methods for the solution of this kind of problems are not available. In the problem considered one simplifying assumption can be made. In practice, we always deal with measurements being discrete in time. This makes the situation easier and permits the construction of an algorithm for finding the solution in a general case of nonstationary random field.

Under the above assumptions, representation (21) is valid for the covariance matrix function of measured characteristics $y(t)$, whereas for the cross-covariance function $z(t)$ and $y(t)$ the formula

(27) $$R_{ZY}(s, \tau) = R_Z(s, \tau)\cdot A^*(\tau)$$

is true.

Now let us consider system (26) in the case of discrete measurements performed at the times $t_1, \ldots, t_N$ ($t_0 \leqq t_1 \leqq \leqq \ldots \leqq t_N \leqq T$). Introduce the following symbols:

(28) $$K_p(s) = K(s, t_p);\ R_Y(p, r) = R_Y(t_p, t_r);\ R_{ZY}(s, r) = R_{ZY}(s, t_r);\ (p, r = 1, \ldots, N).$$

Taking account of (21), (27) and (28) equation (26) is rewritten in the form of a set of N algebraic systems

(29) $$\sum K_p(s)\cdot R_Y(p, r) = R_{ZY}(s, r) \quad (r = 1, \ldots, N)$$

relative to the N unknown matrices $K_p(s)$ ($p=1, \ldots, N$) each of which has the dimension $n\times m$. Thus, the total number of unknown parameters is equal to $N\times n\times m$. Taking account of the fact that the dimension of each matrix $R_Y(p, r)$ and $R_{ZY}(s, r)$ are $m\times m$ and $n\times m$, respectively, we arrive at the conclusion that system (29) consists of $N\times n\times m$ equations. With the increase in the number of simultaneously assimilated arrays of the data of measurements N the possibilities of the immediate numerical inversion (19) decrease. With this, the accuracy of the obtained results diminishes. Speaking about a given concrete physical problem one should bear in mind that modern measuring apparatus of meteorological satellites [4] permits the performance of more than 100 independent measurements of outgoing radiation over comparatively small areas of the earth's surface (400×500 km^2) within which usually exist intense correlations of meteorological elements. Thus, for the actual solution of the problem it is necessary to find the constructive method of the decrease in the dimension of the system of the form (29) permitting one to find the solutions without any serious limitation of the number N.

Below, it will be shown that using the specific structure (29) it is possible to transform this system into the triangle block form, which is equivalent to its inversion. Formally, this is connected with the construction of the canonical basis $\{Y_i(t)\}$ ($i=1, \ldots, N$) for the random function $y(t)$.

We shall transform in succession equations (29). Let us begin with $r=1$. Multiply by $[R_Y(1,1)]^{-1}$ each term of the right and left parts of the first equation (29). Introducing the symbols $Y_1(p) = R_Y(p, 1)\cdot[R_Y(1, 1)]^{-1}$, $C(s, 1) = = R_{ZY}(s, 1)\cdot[R_Y(1, 1)]^{-1}$ we shall write this equation in the form

(30) $$\sum_{p=1}^{N} K_p(s)\cdot Y_1(p) = C(s, 1)$$

It is evident that $Y_1(1) = I$, (I-is the matrix of the identical transformation). Now in equations (29) with the numbers $r=2, 3, \ldots, N$ we perform subtraction of each term from the right and left parts of the following expression:

$$\left(\sum_{p=1}^{N} K_p(s)\cdot Y_1(p)\right)\cdot R_Y(1, r).$$

As a result, we obtain

(31) $$\sum_{p=1}^{N} K_p(s)\cdot R_Y(p, r, 1) = R_{ZY}(s, r, 1) \quad (r = 2, 3, \ldots, N)$$

in which the following symbols are used $R_Y(p, r, 1) = R_Y(p, r) - Y_1(p)\cdot R_Y(1, r)$, $R_{ZY}(s, r, 1) = R_{ZY}(s, r) - R_{ZY}(s, 1)\cdot \cdot R_Y^{-1}(1, 1)\cdot R_Y(1, r)$.

Let us multiply by $[R_Y(2, 2, 1)]^{-1}$ each term of both parts of equation (31). As a result, we obtain

$$\sum_{p=1}^{N} K_p(s) \cdot Y_2(p) = C(s, 2) \tag{32}$$

in which $C(s, 2) = R_{ZY}(s, 2) \cdot [R_Y(2, 2, 1)]^{-1}$. For the matrix function $Y_2(p) = R_Y(p, 2, 1) \cdot [R_Y(2, 2, 1)]^{-1}$ $Y_2(1) = 0$ (0 is zero matrix), $Y_2(1) = I$ are valid.

Subtracting the sum $\left(\sum_{p=1}^{N} K_p(s)\ Y_2(p)\right) \cdot R_Y(2, r, 1)$ from both parts of equation (29) with the numbers $r = 3, \ldots, N$, we shall obtain a new system

$$\sum_{p=1}^{N} K_p(s) \cdot R_Y(p, r, 2) = R_{ZY}(s, r, 2) \tag{33}$$

in which the following symbols are used

$$R_Y(p, r, 2) = R_Y(p, r, 1) - R_Y(p, 2, 1) \cdot R_Y^{-1}(2, 2, 1) \cdot R_Y(2, r, 1)\,,$$

$$R_{ZY}(s, r, 2) = R_{ZY}(s, r, 1) - R_{ZY}(s, 2) \cdot [R_Y(2, 2, 1)]^{-1} \cdot R_Y(2, r, 1)\,.$$

Let us multiply both parts of (33) with the number $r = 3$ by $[R_Y(3, 3, 2)]^{-1}$. Then we obtain

$$\sum_{p=1}^{N} K_p(s) \cdot Y_3(p) = C(s, 3) \tag{34}$$

in which $C(s, 3) = R_{ZY}(s, 3, 2) \cdot [R_Y(3, 3, 2)]^{-1}$,

$$Y_3(p) = R_Y(p, 3, 2) \cdot [R_Y(3, 3, 2)]^{-1}\,.$$

One can easily see that $Y_3(1) = Y_3(2) = 0$, $Y_3(3) = I$. Continuing this process for the subsequent values of r in the above manner and taking account of the properties of the matrix functions $Yr(p)$ forming the canonical basis for $y(t)$ we obtain for the last equation (with the number $r = N$) the following formula:

$$K_N(s) = C(s, N) \tag{35}$$

Passing to the reverse of the computational process we shall the reader remind that for the equation with the number $r = N-1$ the relations $Y_{N-1}(N-1) = I$, $Y_{N-1}(p) = 0$ $(p < N-1)$ are valid. Therefore, taking account of (35) we obtain

$$K_{N-1}(s) = C(s, N-1) - C(s, N) \cdot Y_{N-1}(N)$$

Continuing this process over the diminishing value of the number r and taking account of $Y_r(r) = I$, $Y_r(p) = 0$ $(p < r)$ we find in succession all the matrix operators $K_r(s)$ $(r = 1, \ldots, N)$. The recursive algorithm of the construction of a filter was formulated in [7].

In the following sections of the paper the possibilities of different versions of the method are illustrated in the analysis of temperature and geopotential fields. The procedure of inversion of the matrix R_I having a large dimension was performed using the considered algorithm.

4. First order filters

Before passing to the results of the numerical modelling we shall dwell upon the application of the initial data. For the calculations of statistical structural parameters of the temperature field use was made of the results of radiosounding for summer and winter seasons (June–July 1972, December-January 1971/72, 60 performances in each sample) referring to 13 stations on the European territory of the USSR. The corresponding characteristics of the geopotential field were calculated by formula (15). All estimations were carried out for eight atmospheric layers determined by standard isobaric surfaces: 50, 100, 200, 300, 500, 700, 850, 1000 mb. The relative thermal stability of the atmosphere in summer 1972 resulted in the standard temperature deviations σ_T being equal only to 2 °C–3 °C. Owing to the lack of considerable fronal zones in a summer season it was possible to conduct the statistical analysis under conditions of sufficiently homogeneous fields. In the case of winter statistics the values σ_T were 5 °C–7 °C. The temperature field referring to the winter season was much less homogeneous. The calculation of the elements of the matrix operator A interrelating the variations of the measured radiation characteristics and the fluctuations of temperature distribution was based upon the transmission functions of the real atmosphere in the eight narrow spectral intervals of the 15 µm CO_2 band corresponding to the measuring scheme of the infrared spectrometer SIRS-B [4].

An important element of the investigation was the determination of the degree of the stability of assimilation schemes to the errors in fixing the initial information characterized by the random vector $\varepsilon(x)$. The procedure of modelling the values of the covariance matrix R_ε used in the current paper coincides fully with that exposed in [8]. Results of the calculations are given in the standard form in Table I with the explanations of the symbol abbreviations.

According to (12) there are two types of filters of the first order. With $x_0 = x_1$ (the estimation is conducted at the point of measurements) the case of satellite information is of interest: $A \neq I$. The given type of the filter is reduced to the conventional method of statistical regularization. When $x_0 \neq x_1$ the filter algorithm carries out the optimal extrapolation.

Table 1.

Input Information								Results of statistical analysis				
								Standard deviations of estimates				
								Temperature (°C)		Geopotential (M)		
Assimilation scheme	Number of version	Season	Point of the input of information	Distance from the node (km)	Kind of information	Accuracy a(°C); c (%)	Points of estimation (nodes)	maximum values	mean values	700 mb	500 mb	200 mb
1	2	3	4	5	6	7	8	9	10	11	12	13
S	1	su	3	0	s	2.0	3	1.8	1.2	26	35	47
	2	su	4	0	s	2.0	4	1.7	1.1	25	32	46
	3	su	5	0	s	2.0	5	1.6	1.1	22	30	43
	4	su	9	0	s	2.0	9	2.4	1.5	29	40	59
	5	w	8	0	s	2.0	8	2.6	1.9	27	46	63
	6	w	7	0	s	2	7	2.5	1.8	25	43	64
	7	w	1	0	s	2	1	2.9	2.1	30	49	65
	8	w	5	0	s	2	5	2.4	1.7	26	48	62
	9	w	6	0	s	2	6	2.6	1.8	24	47	60
	10	w	4	0	s	2	4	2.8	1.9	24	45	65
	11	w	11	0	s	2	11	3.6	2.3	32	56	73
	12	w	3	0	s	2	3	3.9	2.1	29	51	66
EX	1	su	1	180	s	1	5	2.1	1.4	28	42	64
	2	su	1	180	s	2	5	2.3	1.5	30	43	66
	3	su	1	180	a	0.8	5	1.7	1.3	22	37	56
	4	w	1	180	s	1.0	5	2.9	2.2	35	65	79
	5	w	1	180	a	0.8	5	2.6	1.8	32	59	72
	6	su	9	480	s	2	5	3.0	1.9	34	51	77
	7	su	9	480	a	0.8	5	2.2	1.5	26	39	62
	8	w	8	540	s	2	5	3.8	2.5	43	78	97
	9	w	8	540	a	0.8	5	2.7	1.9	38	67	74
	10	su	2	800	s	2	5	3.2	1.9	36	49	81
	11	su	13	1100	s	2	5	3.1	2.0	39	54	80
	12	su	4	300	s	2	3	2.1	1.8	29	42	68
	13	w	4	300	s	2	3	3.1	2.5	40	74	98
	14	su	1	660	s	2	3	2.8	2.1	33	51	75
	15	su	8	960	s	2	3	2.9	2.2	37	64	81
	16	su	7	1050	s	2	3	3.0	2.2	43	67	89
	17	su	6	1400	s	2	3	3.0	2.2	47	72	92
	18	su	12	340	s	2	9	3.1	2.0	40	59	94
ins	1	su	2	330	s	1						
			12	340	s	1	9	2.7	1.6	32	49	77
	2	su	2	330	s	2						
			12	340	s	2	9	2.8	1.7	33	52	79
	3	su	2	330	a	0.8						
			12	340	a	0,8	9	2.1	1.3	21	34	47
	4	w	1	190	s	2						
			7	320	s	2	6	2.7	2.0	34	59	84
	5	w	3	680	s	2						
			8	530	s	2	5	3.0	2.2	40	65	94
	6	su	2	330	s	2						
			8	370	s	2						
			12	340	s	2	9	2.6	1.5	30	42	65
	7	w	1	190	s	2						
			7	320	s	2						
			11	330	s	2	5	2.3	1.8	30	51	72
	8	w	3	680	s	2						
			8	530	s	2						
			6	610	s	2	5	2.5	1.9	34	56	74
	9	su	3	300	s	2						
			9	380	s	2						
			12	510	s	2	4	1.9	1.3	19	36	65
	10	su	3	300	a	0.8						
			9	380	a	0.8						
			12	510	s	2	4	1.5	1.1	16	30	43
	11	su	3	300	s	2						
			9	380	s	2						
			12	510	s	2						
			5	420	s	2	4	1.8	1.2	27	36	61
	12	su	2	330	s	2						
			8	370	s	2						
			12	240	s	2						
			1	320	s	2	9	2.6	1.6	28	43	72
	13	w	1	190	s	2						
			7	320	s	2						
			11	330	s	2						
			4	420	s	2	5	2.1	1.7	27	49	68
	14	w	3	680	s	2						

Fortsetzung Table 1

Input Information								Results of statistical analysis				
								Standard deviation of estimates				
								Temperature (°C)		Geopotential (M)		
Assimilation scheme	Number of version	Season	Point of the input of information	Distance from the node (km)	Kind of information	Accuracy a (°C); c (%)	Points of estimation (nodes)	maximum values	mean values	700 mb	500 mb	200 mb
1	2	3	4	5	6	7	8	9	10	11	12	13
			8	530	s	2						
			6	610	s	2						
			11	330	s	2	5	2.4	1.8	30	52	71
as	1	w	3	300	s	2						
			4	0	s	2	4	2.1	1.3	16	28	41
	2	w	3	660	s	2						
			4	350	s	2						
			1	0	s	2	1	1.6	1.2	16	27	35
	3	w	3	1050	s	2						
			4	750	s	2						
			1	390	s	2						
			7	0	s	2	7	1.4	1.0	14	33	31
	4	w	3	1400	s	2						
			4	1100	s	2						
			1	680	s	2						
			7	360	s	2						
			6	0	s	2	6	1.3	1.0	15	24	29
	5	w	3	960	s	2						
			8	0	s	2	8	3.2	1.7	21	28	45
	6	w	3	1050	s	2						
			8	410	s	2						
			7	0	s	2	7	2.6	1.4	16	25	43
	7	w	3	840	s	2						
			8	790	s	2						
			7	560	s	2						
			11	0	s	2	11	2.2	1.2	15	21	33
	8	w	3	680	s	2						
			8	540	s	2						
			7	320	s	2						
			11	330	s	2						
			5	0	s	2	5	2.0	1.0	13	19	29
aas	1	w	4	680	a	0.8						
			8	0	s	2	8	1.8	1.2	17	28	31
	2	w	3	960	s	0.8						
			8	0	s	2	8	1.9	1.3	18	26	37
	3	w	4	760	a	0.8						
			8	410	a	0.8						
			7	0	s	2	7	1.2	0.9	13	17	26
	4	w	4	420	a	0.8						
			8	410	a	0.8						
			7	390	a	0.8						
			1	0	s	2	1	0.9	0.7	10	14	19
aas	5	w	4	420	a	0.8						
			8	540	a	0.8						
			7	320	a	0.8						
			1	180	a	0.8						
			5	0	s	2	5	0.8	0.5	7	10	13
	6	w	3	680	a	0.8						
			8	540	a	0.8						
			7	320	a	0.8						
			11	330	a	0.8						
			5	0	s	2	5	1.0	0.6	9	12	15
ina	1	w	4	420	a	0.8						
			8	410	a	0.8						
			7	390	a	0.8	1	1.7	1.2	18	31	41
	2	w	4	420	a	0.8						
			8	540	a	0.8						
			7	320	a	0.8						
			1	180	a	0.8	5	1.4	0.9	14	27	33

Symbols

1. Numbers of stations: 1 – Moscow, 2 – Voieikovo, 3 – Minsk, 4 – Smolensk, 5 – Ryazan, 6 – Kazan, 7 – Gorky, 8 – Vologda, 9 – Bologoe, 10 – Velikie Luki, 11 – Voronezh, 12 – Pskov, 13 – Kem
2. Kind of the information: a – aerological, s – Satellite.
3. Season: su – summer, w – winter.
4. Assimilation scheme: s – estimate of meteorological elements from the individual satellite measurements (conventional method); ex – extrapolation; ins – interpolation of satellite data; ina – interpolation of aerological information; as – spatial adjustment of satellite data; aas – adjustment of aerological and satellite information.

To estimate the possibilities of the first filter we shall turn to the results of numerical experiments (Table 1, assimilation schemes). In the right part of Table 1 a number of accuracy characteristics is given as obtained on the basis of formulae (14), (16). The mean $\hat{\sigma}_T$ and maximum δ_T values of standard deviations of temperature estimates refer to the eight standard levels mentioned above. Further, standard deviations of geopotential estimates for three isobaric surfaces 700, 500 and 200 mb are presented. In the case of a thermally stable atmosphere (summer statistics) the accuracy of the retrieval of temperature and geopotential profiles is sufficiently high (versions *s*–1–4). The value $\hat{\sigma}_T$ does not exceed 1.5 °C, maximum errors not exceeding 2 °C. The rms errors of the estimate of values of the geopotential *H* (700), *H* (500), *H*(200) are 10–20. 20–30 and 25–50 m, respectively. The analysis of the results under conditions of a less stable atmosphere (winter statistics) indicates a much lower accuracy of the obtained results (*S*–5–11). On the average, the values $\hat{\sigma}_T$ are equal to 2 °C. Maximum errors of the retrieval of the temperature profile in a number of cases exceed the value 3 °C. Mean errors in the determination of geopotential heights *H* (700), *H*(500) and *H* (200) increase up to the values of 20–30, 40–60 and 50–80 m., respectively. Note that the values mentioned constitute 25–35% of the apriori uncertainty of the field characterized by the diagonal elements of the auto-covariance matrix R_H.

The information obtained with the help of the filter considered cannot be used immediately in the numerical forecasting models. Previously it must be interpolated in the nodes of the regular grid. Therefore, it is necessary to estimate the possible losses in accuracy caused by interpolation procedures. First, we shall consider the simplest assimilation scheme – the extrapolation of satellite information from a certain reference point in the node of the grid (scheme ex). It is reasonable to compare the obtained results with the error of the remote sounding performed immediately in this node (schemes). The comparison between the versions (ex – 2, 6, 10, 11) and (s – 3), and between (ex–12, 14–17) and (s–1) shows that the accuracy of extrapolation decreases most noticeably in the distance interval of 200 to 300 km. In this case the mean errors $\hat{\sigma}_T$ increase by 0.4–0.8 °C, the errors in the estimate *H* (700), *H* (500) and *H* (200) increase by 10,20 and 30–40 m, respectively. For a less homogeneous field (winter statistics) the corresponding increments of the error constitute 0.8 °C–1.5 °C (temperature), 10–20, 30–40 and 50–70 m (geopotential). In the distance interval of 300–800 km (ex–6, 10, 11, 12, 14, 13, 8) the accuracy of extrapolation decreases less noticeably. On the average, the error of the temperature estimate increases by 0.2 °C–0.3 °C, and that of the geopotential by 5–15 m. In the extrapolation to the distances exceeding 1000 km (ex – 11, 16, 17) stabilization of the error of estimates is observed, their values, however, approach the level of the apriori ambiguity of the field. The results of numerical experiments indicate that the extrapolation of satellite information just as in the case of aerological (ex–2, 5, 7, 9) leads to its considerable losses.

A two times increase in the level of errors ε leads to the decrease in the accuracy of temperature extrapolation by 0.1 °C–0.2 °C and that of the geopotential by 3–5 m (ex–1, 2). Note, that with distance the discrepancies in the accuracy of the result, as a rule, decrease. The data obtained are indicative of the practical stability of the given assimilation scheme.

Thus, though extrapolation as an assimilation scheme is not of great interest, the analysis made permitted the detection of the regularities connected with a spatial estimation from the data of remote sounding.

5. Optimal interpolation

Statistical interpolation, an important constituent of the modern objective analysis, is of practical interest also in the case of satellite information. The statistical approach used by us combines the interpretation algorithm and the method of spatial analysis Due to this it is possible to transform the data of remote sounding immediately in the estimates of meteorological elements in the nodes of three-dimensional grid.

Let us consider a number of interpolation schemes using the network stations as the points of reference of the data of satellite measurements and nodes (assimilation scheme ins). From the versions (ex–18) and (ins–2,6) it is seen that the successive increase in the order of spatial filter leads to a noticeable increase in accuracy. The filter of the third and second order (ins–6, 9–12) performs interpolation in the node. Thereupon, an accuracy of the obtained estimates is close to that achieved in the interpretation of satellite measurements conducted immediately in the node (s–2,4). In the case of less homogeneous field (winter statistics) the errors of optimal interpretation in the node (ins–7, 8, 13, 14) differs slightly from the error of the immediate remote sounding at this point (s–8). The results referring to the versions (ins–4, 5, 7, 8) permit the detection of a certain dependence of the accuracy of interpolation on the distance from the node of those points at which the information is fixed. Note, that the estimates referring to the predictors of the third and fourth order are in a rather close correspondence with the values of errors (ins–6, 12, 7, 13, 9, 11, 8, 14). The estimates obtained for the version (ins–2, 3, 9, 10) give the possibility to compare the accuracies of interpolation of two kinds of meteorological information. Possessing aerological information it is possible to obtain the estimates of temperature and geopotential much more accurate than on the basis of satellite data. For the assimilation schemes considered the corresponding differences are, on the average, 0.3 °C–0.5 °C and 10–20 m. The results of the versions (ins–1, 2) demonstrate the practical stability of the algorithm of the errors in fixing the initial information.

In the results of numerical experiments the errors of interpolation vary within wide limits. The mean error of temperature estimates is 1.2 °C–2.5 °C, the errors of determination of geopotential for isobaric surfaces of 700, 500 and 200 mb vary within the range of 20–40, 30–60 and 50–90 m, respectively. Thus, though the procedure of optimal interpolation performs the transformation of satellite information in the nodes of the grid practically without losses in accuracy, the level of errors of the obtained estimates exceeds obviously the values acceptable for numerical forecasting models. Hence, for the regions not provided with the aerological information spatial density of satellite measurements, one sounding per 300–500 km, is unsatisfactory. In [8] it is shown that, when the spatial step of satellite measurements is smaller than 100 km, the accuracy of interpolation of temperature estimates increases considerably reaching the values of 0.8 °C–1.5 °C. With this, the errors of determination of geopotential decrease by a factor of 1.5–2. The minimum value of the spatial step with which the remote sounding may be performed is determined, in particular, by spatial resolution of measuring instruments constituting, at present, the value exceeding 30 km. When the density of measurements equal to 30–40 km is achieved the interpolation problem is changed into a problem of spatial adjustment of satellite information.

6. Spatial filtering of the data of remote sounding

The low accuracy of the remote method of determination of temperature and geopotential based upon the conventional interpretation algorithms restricts the possibilities of the application of obtained results to numerical forceasting. Though in recent years a number of different mathematical approaches to the solution of the inverse problem has been proposed the accuracy of the solution remains practically unchanged. The improvement in the accuracy of remote methods may be expected only if more adequate statement of the problem is used. The generalized approach to the solution of inverse problems of satellite meteorology [6, 7] proposed by us implies the use as an apriori information not only of the characteristics of the vertical statistical structure of the atmospheric physical parameters sought but also of the values describing more intensive horizontal correlations of the elements of the meteorological field. Making use of the terms applied in the objective analysis [3] it is possible to state that this method carries out spatial adjustment of the data of remote measurements. Since one can speak of two kinds of meteorological information, two kinds of adjustment are possible. It is reasonable to consider the adjustment between heterogeneous and homogeneous information.

In the analysis of the field from the data of remote sounding of the regions not provided with aerological information an important part must be played by the method of spatial adjustment. We have conducted numerical experiments using the network stations as supporting points (assimilation scheme – as.) The adjustment of the pair of satellite measurements with a distance from each other equal to 300 km (version as–1) permits the increase in the accuracy of temperature estimation by 0.5 °C, and that of the geopotential by 10–20 km as compared to the conventional interpretation method (s–10). The comparison with the version (as–5) permits the detection of the dependence of the effectiveness of adjustment on the distance between the supporting points. The use of the third order filter (as–2,6) leads to the increase in the accuracy of estimation of both meteorological elements, on an average, by a factor of 1.5 as compared with the possibilities of the statistical regularization method (s–6,7). The versions (as–7,8) demonstrate a successive increase in the accuracy of the estimates with the increase in the order of the filter. The adjustment of five satellite measurements at the distance from one another equal to 300–600 km permits the achievement of the level of the mean accuracy of temperature estimation equal to 1 °C. The errors of the determination of the value of the geopotential H (700), H (500), H (200) are 15–30 and 20–40 km, respectively. Versions (as–3,4) permit one to state that at the distance between the supporting point comparable with the values of the correlation radius of the meteorological fields (i.e. above 1000 km) sufficient of accuracy is observed with the adjustment of four measurements.

The most effective are the filters of high orders referring to a dense network of supporting points. According to the results [8] the accuracy of temperature estimation may reach the level of 0.5 °C with the spatial step of measurements less than 100 km. Yet, the value of the order of the used filter is restricted to the possibilities of electronic computers. Therefore, the problem of the determination of the optimal relation between the density of supporting points and the order of the filter becomes very urgent.

7. Adjustment of satellite and aerological information

In the case of remote sounding of the regions partially provided with the information, the adjustment of both kinds of meteorological information is of certain importance. The corresponding calculations have been conducted for several situations (assimilation scheme – 11s). From the version (aas–1) it is seen that the application of aerological information from the station at a distance of 680 km ensures the increase in the accuracy of estimation of the geopotential by 10–30 m (s–5). The effectiveness of the given kind of adjustment depends also on the distance between the supporting points (aas–1,2). The increase in the number of supporting stations leads to the considerable increase in the effectiveness of remote method. With the presence of the information from 2–3 supporting stations situated within the radius of 400–800 km the estimation error at the point of measurement from a satellite is less than 1 °C for temperature and no more than 30 m for the geopotential (aas – 3,4). The adjustment of satellite information to the data of more extensive meteorological network leads to a much more appreciable increase in the effectiveness of the analysis. The accuracy of temperature estimation can reach the value of 0.5 °C, and that of the geopotential 10–15 m (aas–4–6) even with the mean distance of supporting stations (300–500 km).

Since at the subsatellite points the estimation of the meteorological elements can be performed on the basis of the interpolation of aerological information from supporting stations it is reasonable to estimate the immediate contribution of remote sounding data to the final result. For this purposes, the procedure of spatial interpolation of aerological data in the node of the gridd was performed (assimilation scheme – ina), for the latter the point of the reference of satellite information was chosen. In this sense, versions ina–1,2 correspond to the conditions of calculations for aas–4,5. The comparison of the obtained results gives a certain notion about the value of satellite information in the analysis of the field. The consideration of the independent data of remote sounding ensures the specification of temperature estimates by 0.3 °C–0.5 °C, and that of the geopotential by 10–20 m, which is 30–40% of the level of the errors of interpolation of aerological information in the node. Thus in the presence of the meteorological network of stations with the mean density of (300–500 km) satellite data if used correctly may contribute considerably to the objective analysis of the meteorological field.

In conclusion, we shall make some remarks concerning the estimates presented in the paper. The method of numerical modelling used cannot provide definitive answers about the possibilities of the proposed statistical approach. However, as shown above, using this approach it is possible to detect not only the intervals of the expected errors of the estimates of meteorological elements but to point out the main prerequisites for implementing the optimal interpretation of spatial satellite information and its adjustment to the available aerological data.

It should be noted that the limits of errors in the estimation of the geopotential given in the paper must be considered as slightly underestimated due to the non-consideration of the errors in fixing surface pressure.

Contrary to the aerological information, the accuracy of the remote sounding method depends not only on measurement errors but also on the random variations of optical parameters of the atmosphere due to the uncontrolled fluctuations of the content of its absorbing components (water vapor, aerosol, etc.). The lack of the information on spatial statistical structure of these variations may, in a certain extent, effect the accuracy of a given method. Therefore, the stability of the spatial assimilation schemes considered regarding the errors of the input data revealed during the numerical experiments is the property of great practical importance.

References

[1] *Anderson T.*, An Introduction to Multivariate Statistical Analysis. 1960, New York. John Wiley and Sons. 500 p.
[2] *Baumhefner D., Yulian P.*, The structure and growth of error introduced by an observational system of remotely sensed temperature profiles.
[3] *Gandin L. S.* Objective Statistical Analysis of Meteorological Fields. L., Gidrometeoizdat, 1963, 285 p.
[4] *Houghton J. T., Taylor F. W.*, Remote sounding from satellites and space probes of the atmosphere of the earth and planets. Rep. Prog. Phys., 1973, v. 36, p. 827–919.
[5] *P. Morel, Bengtsson L.*, The performance of space observing systems for the first GARP global experiment. The GARP programme on numerical experimentation. Report No. 6, July 1974, 31 p.
[6] *Pokrovsky O. M.*, Assimilation of the Data of Direct and Indirect Measurements in the Statistical Analysis of Meteorological Fields. „Meteorology and Hydrology“, 1974, No. 6, pp. 33–39.
[7] *Pokrovsky O. M.*, Optimal Statistical Procedures of Space-Time Mastering of Indirect Meteorological Information. „Meteorology and Hydrology“, 1975, No. 7, pp. 21–29.
[8] *Pokrovsky O. M.* and *Ivanykin E. E.*, Estimation of the Possibilities of Spatial Assimilation of the Data of Direct and Indirect Measurements of Temperature Field. „Meteorology and Hydrology“, 1976, No. 3, pp. 27–36.
[9] *Sherman S.*, Non Mean-Square Error-criteria. Trans. IRE IT-4, 1958, pp. 125–136.
[10] *Smith W., Fritz S.*, On the statistical relation between geopotential height and temperature pressure profiles. Techn. Rep. NESTM-18, Nov. 1969, 42 pp.

Address: O. M. Pokrovskij and E. E. Ivanykin, Department of Atmospheric Sciences, Physical Institute Leningrad, University, UdSSR Leningrad-Petrodovrets 98904

551.509.313
551.509.317

Zur statistisch optimalen Kombination von aerologischen und SIRS-Daten bei der numerischen Geopotentialanalyse

Von **J. Kluge** und **M. Petzold**

Mit 3 Abbildungen

Zusammenfassung: Unter Verwendung der Methode der optimalen Interpolation werden aerologische Geopotentialdaten und durch indirekte Sondierungen bestimmte Geopotentialdaten unter Beachtung der verschiedenartigen Fehlerstruktur der Informationsarten optimal kombiniert. Im Rahmen einer Fallstudie werden mittels einer numerischen Einflächenanalyse die Wirkung des im Vergleich zur aerologischen Information vergrößerten Fehlers der SIRS-Daten (F^s-Effekt), der Korrelation dieser Fehler (K-Effekt) und der Zeitverschiebung (τ-Effekt) untersucht. Diese Experimente werden durch a priori Abschätzungen für eine Einpunkt-Analyse ergänzt. F^s-Effekt und K-Effekt beeinflussen das Analyseresultat wesentlich.

Summary: Applying the method of optimal interpolation, aerological geopotential data and remote sensing-geopotential data are combined taking into account the different error structure of the data. In a case study the effect of the greater satellite data error (F^s-effect), the error correlation (K-effect) and the influence of time difference (τ-effect) are studied. The real data experiment is completed by a-priori estimates on the basis of a one-point analysis. The F^s and K-effects influence the analysis significantly.

1. Einleitung

Werden bei der numerischen Analyse des Geopotentials nur die konventionellen aerologischen Daten verwendet, dann vereinfacht sich das Analyseschema dadurch beträchtlich, daß die Fehler der aerologischen Daten, nachdem Übertragungs- und Kodierungsfehler eliminiert wurden, als weißer Lärm behandelt werden dürfen. Die für die numerische Analyse folgenreichste Besonderheit der durch indirekte Sondierung bestimmten Temperatur- bzw. Geopotentialdaten (im folgenden kurz mit dem nicht ganz zutreffenden, aber populären Namen SIRS-Daten bezeichnet) besteht darin, daß sie eine sehr komplizierte Fehlerstruktur besitzen, die den weißen Lärm-Ansatz unannehmbar macht. Für diese veränderte Fehlerstruktur sind im wesentlichen drei Gründe maßgebend. Erstens ist bei der indirekten Sondierung der Maßstab der räumlichen Mittelung größer als bei einem aerologischen Aufstieg. Zweitens erzeugen physikalische, numerische und apparative Faktoren bei diesem unter sehr komplizierten Bedingungen arbeitenden, höchst komplexen Sondierungsverfahren vielfältige Ungenauigkeiten. Drittens wird, während bei der aerologischen Sondierung für jeden Aufstieg ein neues Gerät verwendet wird, vom Satelliten mit ein und demselben Gerät über lange Zeit gearbeitet. Der erste Grund führt vor allem zur Minderung der Auflösung in der Vertikalen, der zweite zur Erhöhung des Fehlerniveaus allgemein und der dritte zu beträchtlicher interner Fehlerkorrelation. Zur komplizierten Fehlerstruktur der SIRS-Daten kommt hinzu, daß sie asynoptisch sind.

In einer Reihe von Arbeiten [6], [7], [10] und [8] wurden diese Besonderheiten der SIRS-Daten im Hinblick auf die numerische Analyse untersucht, allerdings nur im Rahmen von a priori Abschätzungen ohne numerische Analyse realer Daten. Deshalb sah sich die GARP-Arbeitsgruppe für numerische Experimente [9] veranlaßt zu empfehlen: „. . . die Datenassimilationsverfahren so zu gestalten, daß derartige Fehlerstrukturen einbezogen werden können . . ."

Unter Beachtung dieser Empfehlung stellen wir uns in dieser Untersuchung die Aufgabe, reale aerologische Geopotential- und SIRS-Daten im Rahmen einer numerischen Einflächenanalyse statistisch optimal zu kombinieren und dabei die Fehlerstruktur explizit einzuführen. Die Untersuchung ist als Fallstudie angelegt, bei der insbesondere die Größenordnung der einzelnen Effekte anhand realer Analysen abgeschätzt werden soll. Diese Abschätzungen werden ergänzt durch Berechnung des theoretischen Interpolationsfehlers und der Interpolationsgewichte für simulierte Daten.

2. Interpolationsformel

Als Interpolationsformel verwenden wir bei der numerischen Analyse die in der Arbeit [8] vorgeschlagene, spezialisiert für den Fall, daß aerologische Geopotentialinformation und Satellitendaten vorliegen. Die entsprechenden Formeln werden nachstehend angeführt.

Der Schätzwert $\tilde{f}_{00}$ des Geopotentials H im Gitterpunkt wird als Linearkombination aus den umliegenden Meldungen $\tilde{f}_{ki}$ bestimmt

$$\tilde{f}_{00} = \sum_{k=1}^{m} \sum_{i=1}^{n_k} p_{ki} \tilde{f}_{ki}, \tag{1}$$

($\tilde{f}_{00}$, $\tilde{f}_{ki}$ stellen Abweichungen vom klimatologischen Mittelwert dar, k bezeichnet die Informationsart (einschließlich des Zeitpunktes), i die Position. Wahrer Wert f_{ki} und Fehler Δ_{ki} ergeben den beobachteten Wert $\tilde{f}_{ki} = f_{ki} + \Delta_{ki}$).

Aus der Minimalbedingung $E^2 = \overline{(\tilde{f}_{00} - f_{00})^2} = \text{Min}$ ergibt sich das System zur Bestimmung der Interpolationsgewichte p_{ki}

$$\sum_{l=1}^{m} \sum_{j=1}^{n_l} p_{ki} \overline{\tilde{f}_{ki} \tilde{f}_{lj}} = \overline{f_{00} \tilde{f}_{ki}} \quad k=1 \ldots m, \quad i=1 \ldots n_k \tag{2}$$

bzw. eine Formel zur a priori Abschätzung des Interpolationsfehlers

$$E^2 = \overline{f_{00}^2} - \sum_{k=1}^{m} \sum_{i=1}^{n_k} p_{ki} \overline{f_{00} \tilde{f}_{ki}}. \tag{3}$$

Wir bezeichnen die aerologische Geopotentialinformation mit $\tilde{f}_{1i} = \tilde{H}_i$, die SIRS-Daten mit $\tilde{f}_{2i} = \tilde{S}_i$, falls keine Zeitverschiebung zum Analysetermin vorliegt, und mit $\tilde{f}_{3i} = S(\tau)_i$, falls sie τ [Std.] beträgt. Die entsprechenden Fehlerausdrücke lauten $\overline{\Delta_{1i}^2}/\overline{H^2} = F_i^H$, $\overline{\Delta_{2i}^2}/\overline{H^2} = \overline{\Delta_{3i}^2}/\overline{H^2} = F_i^s$. Die entsprechenden Korrelationsmomente approximieren wir in Übereinstimmung mit [8] wie in Tab. 1 angegeben. Den nach Gl. (3) bestimmten normierten theoretischen Interpolationsfehler $E^2/\overline{H^2}$ bezeichnen wir mit ε.

Tabelle 1

Approximation der statistischen Struktur des Systems

	$\tilde{H}_i$	$\tilde{S}_i$	$S(\tau)_i$
$\tilde{H}_j$	$\overline{H^2}[(1+\varrho/L)\times \times\exp(-\varrho/L) + +\delta_{ij} F_i^H]$	$\overline{H^2}[(1+\varrho/L)\times \times\exp(-\varrho/L)]$	$\overline{H^2}[(1+\varrho^*)\times \times\exp(-\varrho^*)]$
$\tilde{S}_j$		$\overline{H^2}[(1+\varrho/L)\times \times\exp(-\varrho/L) +\mathrm{K}(\varrho)\cdot F_i^s]$	$\overline{H^2}[(1+\varrho^*)\times \times\exp(-\varrho^*)+ +K(\varrho^*) F_i^s]$
$S(\tau)_j$	.		$\overline{H^2}[(1+\varrho/L)\times \times\exp(-\varrho/L)+ +\mathrm{K}(\varrho)\cdot F_s^i]$

$\varrho^* = \sqrt{\varrho^2/L^2 + T^2/\tau^2}$, $K(\varrho)$ siehe Abschn. 5, δ_{ij} – Kroneckersymbol

In die Interpolationsformel gehen demnach eine Reihe von Parametern ein, die in Abhängigkeit von den Eigenschaften der verwendeten Beobachtungsdaten zu aktualisieren sind, insbesondere hinsichtlich der Fehlerstruktur.

3. Testmaterial und Anlage der numerischen Experimente

Der Einfluß der Fehlerstruktur und der Zeitdifferenz auf das Analyseergebnis werden in den Abschnitten 4–6 erörtert. Die Untersuchung erfolgt einmal durch Interpretation des nach Gl. (3) berechneten theoretischen Interpolationsfehlers und der nach Gl. (2) bestimmten Interpolationsgewichte für simulierte Daten. Das jeweilige Informationsangebot, das bei der Interpolation in dem Zentralpunkt benutzt wird, ist auf Abb. 2 angegeben. Zum anderen wurden anhand zweier Datensätze mit einem Einflächenanalyseschema, welches den in Abschn. 2 beschriebenen Interpolationsblock enthält und die Stationssuche nach *Bagrov* u. M. [1] vornimmt, numerische Analysen der AT 500 erzeugt (die Daten wurden vorher manuell kontrolliert). Diese Analysen wurden untereinander und mit der operativ durch den Synoptiker erzeugten Handanalyse, die neben Temp-Daten (Geopotential und Wind) die SIRS-Daten berücksichtigt, verglichen. Für das geographische Gebiet, in dem die SIRS-Daten liegen (Nordatlantik), war kein geeignetes anderes Vergleichsfeld verfügbar. Ein Vergleich der subjektiven Analyse mit einer numerischen Analyse, bei der nur aerologische Geopotentialdaten verwendet wurden (AG genannt), brachte große Unterschiede im Bereich des SIRS-Daten Einflußgebietes (16 gpdam bzw. 22 gdpam in den von uns betrachteten Situationen, s. u.). Auch außerhalb dieses Gebietes, besonders an den Kartenrändern, traten noch relativ große Abweichungen auf, die allerdings nicht auf die im weiteren beschriebene verschiedenartige Einbeziehung der SIRS-Daten reagieren. Demnach kann durch entsprechenden Vergleich mit der Handanalyse der SIRS-Dateneffekt relativ gut herausgeschält werden. Die Bewertung des Informationswertes der SIRS-Daten sowie der Analysetechnik über die Verifikation von Prognosen bleibt einer künftigen Untersuchung vorbehalten. Sehr interessant wäre es auch, die Einbeziehung der SIRS-Daten über Europa zu studieren und dabei durch Aussonderung das aerologische Netz über dem Nordatlantik zu simulieren und als Vergleichsfeld unter Verwendung aller aerologischen Sondierungen berechnete numerische Analyse zu benutzen. Leider waren entsprechende Daten nicht verfügbar.

Verwendet wurden die Situationen vom 30. 04. 75 und 1. 05. 75 00.00 GMT. Die ausgewählten Datensätze entsprechen einem durchschnittlichen operativen Datenangebot. Auf Abb. 1 ist zur Illustration die Handanalyse des Synoptikers nebst dem SIRS-Datenangebot dargestellt. Die SIRS-Werte der AT 500 wurden aus den operativ übermittelten RETOP-Werten durch Addition auf die aus der Bodenanalyse abgeleitete AT 1000 gewonnen, ohne dabei die zeitlichen Unterschiede τ zu berücksichtigen. Speziell anhand dieser Stichproben bestimmte Parameter, die in die Interpolationsformel eingehen, lagen nicht vor.

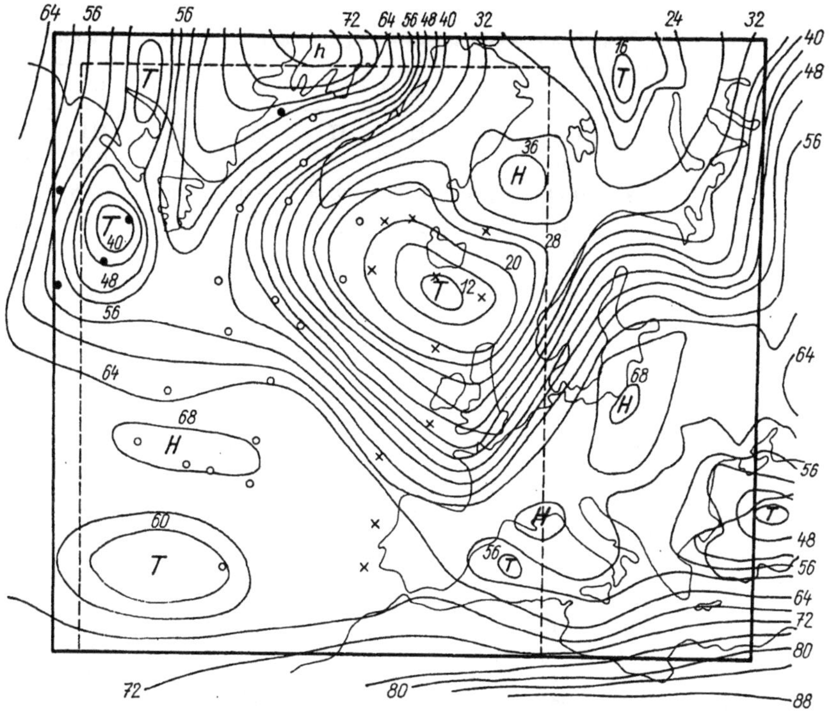

Abb. 1. Handanalyse der AT 500 vom 30. 4. 1975 00.00 GMT. Die Zeichen × ($\tau = 4$ Std.), ● ($\tau = 2$ Std.) und o ($\tau = 0$) bezeichnen die Position der verwendeten SIRS-Meldungen. Das gestrichelt umrandete Gebiet wird auf Abb. 3 gezeigt.

4. Behandlung der SIRS-Daten als aerologische Information (F^s-Effekt)

Für durch Temp-Aufstiege gewonnene Geopotentialdaten wird die Fehlergröße F_i^H überwiegend mit 0,02 angegeben und als zeitlich und räumlich konstant sowie von der synoptischen Situation unabhängig betrachtet. Tatsächlich besteht eine relativ geringe Abhängigkeit vom Sondentyp, die nur bei dem in der ČSSR von *M. Skoda* [11] entwikkelten Analyseverfahren berücksichtigt wird. Für SIRS-Daten ist die entsprechende Fehlergröße generell höher. Jüngste Untersuchungen [2–4], [12] erbrachten den Nachweis, daß der SIRS-Fehler sowohl von der geographischen Breite und dem Höhenniveau als auch der Jahreszeit und der synoptischen Situation sowie dem Gerätetyp abhängt. Den genannten Arbeiten ist zu entnehmen, daß die Fehlergröße Δ_{2i}^2/H^2 für mittlere Breiten und das 500 mbar-Niveau über alle synoptischen Situationen jahreszeitlich etwa zwischen 0,03 und 0,14 schwankt. Sicher wird in Zonen dichter Bewölkung der Wert 0,14 überschritten werden können.

Um nun den Effekt der Erhöhung der Fehlergröße separat untersuchen zu können, behandelten wir die SIRS-Daten als aerologische Information mit erhöhtem Fehler, d. h. Zeitdifferenz τ und Fehlerkorrelation wurden nicht berücksichtigt. Danach reduziert sich die Korrelationsmatrix auf Elemente des Typs $\overline{\tilde{H}_i\tilde{H}_j}$, in der Diagonale stehen die entsprechenden Fehlergrößen. Für die aerologische Information verwendeten wir $F_i^H = 0{,}02$; für SIRS-Daten die Werte 0,02; 0,08; 0,11 und 0,14. Für L wird im folgenden immer $L = 1000$ km gesetzt (s. [8]). Bei der Auswahl der beeinflussenden Stationen werden insgesamt 6 Meldungen gesucht, ohne auf einen bestimmten Anteil von aerologischen oder SIRS-Meldungen zu sehen. Bei den verwendeten Datensätzen gibt es danach über dem Nordatlantik einige Analysepunkte, für die die Interpolation nur auf der Basis von SIRS-Daten vorgenommen wird.

Nehmen wir zunächst eine a priori Abschätzung anhand des theoretischen Interpolationsfehlers und der Interpolationsgewichte vor. Wir denken uns dazu aerologische Geopotentialinformation in der auf Abb. 2 bezeichneten Weise vorgegeben und interpolieren in den Zentralpunkt. Der theoretische Interpolationsfehler nach Gl. (3) und die Interpolationsgewichte nach Gl. (2) sind in Tab. 2 als Funktion von F_i^H und dem Abstand d der Information vom Zentralpunkt dargestellt. Die unter Spalten 1–4 in Tab. 2 angeführten Ergebnisse lassen folgende Schlußfolgerungen zu.

Werden für *alle* betrachteten Informationen (Prediktoren) die Fehlergrößen *gleichmäßig* erhöht, etwa von 0,02 auf 0,08, so ergeben sich keine nennenswerten Änderungen der Interpolationsgewichte.

Der durch den theoretischen Interpolationsfehler abschätzbare Informationswert dagegen sinkt beträchtlich.

Falls jedoch die an der Interpolation beteiligten Prediktoren unterschiedliche Fehlergrößen besitzen, ergeben sich stark unterschiedliche Interpolationsgewichte. Der Beitrag eines Prediktors mit höherem Fehler verringert sich beträchtlich, sein Einflußradius verkürzt sich in Richtung eines Prediktors mit geringerem Fehler.

Die Auswertung der Experimente mit den beiden Datensätzen führt zu folgenden Ergebnissen.

Die starken Abweichungen der nur auf aerologischen Daten basierenden numerischen Analyse (AG) von der Handanalyse (s. Tab. 3) werden bei Einbeziehung der SIRS-Daten in der beschriebenen Weise für alle verwendeten Fehlerwerte beträchtlich verringert.

Tabelle 2

Theoretischer Interpolationsfehler $E^2/\overline{H}^2$ und Interpolationsgewichte p_i als Funktion des Informationsangebotes

Spalten-Nr.	1	2	3	4	5	6	7	8	9	10	11	12
Informationsverteilung	I	I	III	III	II	II	II	II	II	III	III	III
F_i^H	0,02	0,08	0,02	0,02	–	–	–	–	–	0,02	0,02	0,02
F_i^s	–	–	0,08	0,14	0,08	0,08	0,08	0,08	0,08	0,08	0,08	0,08
$K_i(\varrho)$	–	–	δ_{ij}	δ_{ij}	K_1	K_3	K_4	K_2	K_5	K_1	K_3	K_4
τ [Std]	–	–	0	0	0	0	0	0	0	0	0	0
d [km] 50 — $\overline{E^2}/\overline{H^2}$	.005	.020	.008	.009	.067	.074	.073	.074	.070	.009	.009	.009
p_1	.25	.24	.10	.06	.23	.23	.23	.23	.23	.06	.06	.06
p_2	.25	.24	.40	.43	.23	.23	.23	.23	.23	.44	.44	.44
100	.005	.020	.008	.009	.062	.074	.069	.074	.067	.009	.009	.009
	.25	.25	.10	.06	.24	.23	.23	.23	.23	.07	.06	.06
	.25	.25	.40	.43	.24	.23	.23	.23	.23	.43	.44	.44
200	.008	.023	.011	.012	.056	.074	.057	.077	.064	.012	.013	.012
	.25	.25	.11	.07	.24	.24	.24	.23	.24	.08	.07	.09
	.25	.25	.39	.43	.24	.24	.24	.23	.24	.42	.43	.42
300	0.14	.029	.018	.019	.055	.075	.049	.082	.066	.019	.019	.018
	.25	.25	.24	.09	.24	.24	.25	.24	.24	.11	.10	.12
	.25	.25	.37	.42	.24	.24	.25	.24	.24	.40	.41	.39
400	023	.039	.029	.031	.059	.079	.049	.091	.072	.029	.03	.029
	.26	.26	.16	.012	.25	.24	.25	.24	.25	.14	.13	.16
	.26	.26	.36	.040	.25	.24	.25	.24	.25	.37	.39	.36
500	.037	.054	.043	.046	.069	.087	.057	.103	.082	.044	.045	.043
	.27	.26	.19	.14	.26	.25	.26	.25	.25	.17	.16	.18
	.27	.26	.34	.38	.26	.25	.26	.25	.25	.35	.36	.34
600	.055	.072	.062	.066	.085	.099	.074	.120	.098	.063	.064	.062
	.27	0.26	.21	.17	.26	.26	.26	.25	.26	.20	.19	.21
	.27	0.26	.33	.37	.26	.26	.26	.25	.26	.34	.34	.33
700	.078	.095	.085	.090	.105	.115	.096	.140	.117	.086	.086	.085
	.27	.27	.22	.29	.27	.26	.27	.25	.26	.22	.21	.22
	.27	.27	.32	.35	.27	.26	.27	.25	.26	.32	.33	.32
800	.104	.122	.112	.117	.130	.137	.122	.165	.141	.112	.112	.112
	.28	0.27	.23	.20	.27	.27	.27	.26	.27	.23	.23	.23
	.28	0.27	.32	.34	.27	.27	.27	.26	.27	.32	.32	.32
900	.134	.152	.142	.148	.158	.163	.152	.192	.169	.142	.142	.142
	.28	.27	.24	.21	.27	.27	.27	.26	.27	.24	.24	.24
	.28	.27	.31	.34	.27	.27	.27	.26	.27	.31	.31	.31
1000	.167	.186	.176	.182	.191	.195	.186	.223	.201	.176	.176	.176
	.28	.28	.25	.22	.28	.27	.28	.26	.27	.25	.25	.25
	.28	.28	.31	.34	.28	.27	.28	.26	.27	.31	.31	.31

Spalten-Nr.	13	14	15	16	17	18	19	20	21	22	23	24
Informationsverteilung	III	IV	IV	IV	IV	V	V	V	V	V	V	V
F_i^H	0,02	0,02	0,02	0,02	0,02	–	–	–	–	–	–	–
F_i^s	0,08	0,08	0,08	0,08	0,08	0,08	0,08	0,08	0,08	0,08	0,08	0,08
$K_i(\varrho)$	K_2	K_1	K_3	K_4	K_2	δ_{ij}	δ_{ij}	δ_{ij}	K_1	K_1	K_1	K_2
τ [Std]	0	0	0	0	0	3	6	12	3	6	12	3
d [km] 50 — $\overline{E^2}/\overline{H^2}$	.009	.009	.009	.009	.009	.029	.053	.136	.076	.099	.178	.083
p_1	.06	.07	.12	.10	.13	.24	.24	.23	.23	.23	.22	.23
p_2	.44	.04	–.01	.01	–.02	.24	.24	.23	.23	.23	.22	.23
100	.009	.009	.009	.009	.009	.029	.053	.135	.070	.094	.172	.083
	.06	.08	.14	.10	.15	.24	.24	.23	.24	.23	.22	.23
	.44	.03	.04	.00	–.05	.24	.24	.23	.24	.23	.22	.23
200	.013	.013	.013	.013	.013	.031	.054	.133	.063	.085	.162	.084
	.07	.10	.19	.12	.22	.25	.24	.23	.24	.24	.23	.23
	.44	.00	.10	–.02	–.12	.25	.24	.23	.24	.24	.23	.23
300	.020	.021	.020	.021	.020	.037	.058	.134	.062	.083	.156	.089
	.09	.13	.22	.14	.25	.25	.25	.24	.24	.24	.23	.24
	.42	–.03	–.13	–.04	–.17	.25	.25	.24	.24	.24	.23	.24
400	.031	.033	.032	.033	.032	.046	.066	.137	.066	.085	.155	.097
	.12	.16	.24	.26	.27	.26	.25	.24	.25	.25	.24	.24
	.40	–.06	–.14	–.06	–.18	.26	.25	.24	.25	.25	.24	.24
500	.046	.051	.050	.051	.049	.060	.078	.145	.076	.094	.160	.109
	.14	.19	.25	.18	.28	.26	.26	.25	.26	.25	.24	.25
	.38	–.08	–.15	–.07	–.18	.26	.26	.25	.26	.25	.24	.25
600	.066	.074	.072	.074	.071	.078	.095	.150	.107	.107	.169	.125
	.17	.21	.26	.20	.29	.26	.26	.25	.26	.26	.25	.25
	.36	–.09	–.14	–.08	–.18	.26	.26	.25	.26	.26	.25	.25

Fortsetzung von Tab. 2

Spalten-Nr.	13	14	15	16	17	18	19	20	21	22	23	24
Informationsverteilung	III	IV	IV	IV	IV	V	V	V	V	V	V	V
F_i^H	0,02	0,02	0,02	0,02	0,02	–	–	–	–	–	–	–
F_i^s	0,08	0,08	0,08	0,08	0,08	0,08	0,08	0,08	0,08	0,08	0,08	0,08
$K_i(\varrho)$	K_2	K_1	K_3	K_4	K_2	δ_{ij}	δ_{ij}	δ_{ij}	K_1	K_1	K_1	K_2
τ [Std]	0	0	0	0	0	3	6	12	3	6	12	3
700	.089	.101	.100	.102	.098	.101	.116	.174	.110	.126	.183	.145
	.19	.23	.26	.22	.29	.27	.26	.26	.26	.26	.25	.25
	.35	−.10	−.14	−.08	−.17	.27	.26	.26	.26	.26	.25	.25
800	.116	.133	.132	.134	.130	.127	.141	.195	.134	.149	.202	.169
	.21	.25	.27	.24	.30	.27	.27	.26	.26	.27	.26	.26
	.34	−.10	−.13	−.09	−.16	.27	.27	.26	.27	.27	.26	.26
900	.146	.169	.168	.170	.166	.157	.170	.219	.163	.176	.225	.196
	.22	.26	.27	.25	.30	.27	.27	.26	.27	.27	.26	.26
	.33	−.10	−.12	−.09	−.15	.27	.27	.26	.27	.27	.26	.26
1000	.179	.209	.208	.209	.206	.190	.202	.247	.195	.206	.252	.227
	.24	.27	.28	.26	.30	.28	.27	.27	.26	.27	.26	.26
	.32	−.10	−.11	−.09	−.14	.28	.27	.27	.27	.27	.26	.26

Spalten-Nr.	25	26	27	28	29	30	31	32	33	34	35
Informationsverteilung	V	V	VI	VI	VI	VI	VI	VI	VI	VI	VI
F_i^H	–	–	0,02	0,02	0,02	0,02	0,02	0,02	0,02	0,02	0,02
F_i^s	0,08	0,08	0,08	0,08	0,08	0,08	0,08	0,08	0,08	0,08	0,08
$K_i(\varrho)$	K_2	K_2	δ_{ij}	δ_{ij}	δ_{ij}	K_1	K_1	K_1	K_2	K_2	K_2
τ [Std]	6	12	3	6	12	3	6	12	3	6	12
d [km] \ $\overline{E^2}/\overline{H^2}$, p_1, p_2 — 50	.106	.184	.008	.009	.010	.009	.010	.009	.009	.009	.010
	.23	.22	.08	.06	.03	.05	.04	.02	.05	.04	.02
	.23	.22	.41	.44	.47	.44	.46	.48	.45	.46	.48
100	.106	.183	.009	.009	.010	.009	.010	.010	.010	.010	.010
	.23	.22	.09	.06	.03	.06	.04	.02	.05	.04	.02
	.23	.22	.41	.44	.47	.44	.45	.48	.45	.46	.48
200	.106	.181	.012	.013	.014	.012	.013	.014	.013	.013	.014
	.23	.22	.10	.07	.03	.07	.06	.03	.06	.05	.02
	.23	.22	.40	.44	.47	.43	.45	.48	.44	.46	.48
300	.100	.181	.010	.020	.022	.010	.020	.022	.020	.021	.022
	.24	.23	.12	.09	.04	.10	.08	.04	.08	0.6	.03
	.24	.23	.39	.42	.47	.41	.43	.47	.43	.44	.48
400	.116	.184	.029	.031	.034	.030	.032	.034	.031	.032	.035
	.24	.23	.15	.12	.06	.13	.11	.06	.11	.09	.05
	.24	.23	.37	.40	.46	.38	.41	.46	.41	.43	.46
500	.127	.190	.044	.046	.051	.045	.047	.052	.047	.049	.052
	.24	.23	.28	.15	.09	.16	.14	.09	.14	.12	.08
	.24	.23	.35	.38	.44	.36	.38	.44	.39	.40	.45
600	.151	.201	.063	.066	.073	.064	.066	.073	.067	.069	.074
	.25	.24	.20	.18	.12	.19	.17	.12	.16	.15	.10
	.25	.24	.34	.36	.41	.34	.36	.42	.37	.38	.43
700	.160	.215	.085	.089	.097	.087	.090	.098	.090	.092	.099
	.25	.24	.22	.20	.15	.21	.19	.15	.19	.17	.13
	.25	.24	.33	.34	.39	.33	.35	.40	.35	.37	.41
800	.183	.234	.113	.116	.125	.113	.116	.126	.117	.119	.128
	.26	.25	.23	.22	.17	.23	.21	.17	.21	.20	.16
	.26	.25	.32	.33	.38	.32	.34	.38	.34	.35	.39
900	.209	.256	.143	.146	.156	.144	.147	.157	.147	.150	.159
	.26	.25	.24	.23	.19	.24	.23	.19	.22	.21	.37
	.26	.25	.32	.33	.36	.32	.33	.36	.33	.34	.18
1000	.238	.282	.177	.180	.190	.177	.180	.190	.180	.183	.193
	.26	.25	.25	.24	.21	.25	.24	.21	.23	.22	.20
	.26	.25	.31	.32	.35	.31	.32	.35	.33	.33	.36

Die entsprechenden über das gesamte Analysegebiet bestimmten mittleren Abweichungen verringern sich um rund 25% (s. Tab. 3). Betrugen beispielsweise die relativen Maxima der absoluten Abweichungen im SIRS-Einflußgebiet für den 30. 04. 75 bei $F^s = 0{,}08$ 12, 16 und 12 gpdam, so verringerten sich diese bei Einbeziehung der SIRS-Daten auf entsprechend 1, 7 und 8 gpdam (für den 01. 05. 75 lauten entsprechende Zahlen 18, 13 und 24 bzw. 5, 4 und 7 gpdam). Diese Annäherung an die subjektive Analyse demonstriert den hohen Informationswert, der den SIRS-Daten über dem Nordatlantik vom Synoptiker beigemessen wird.

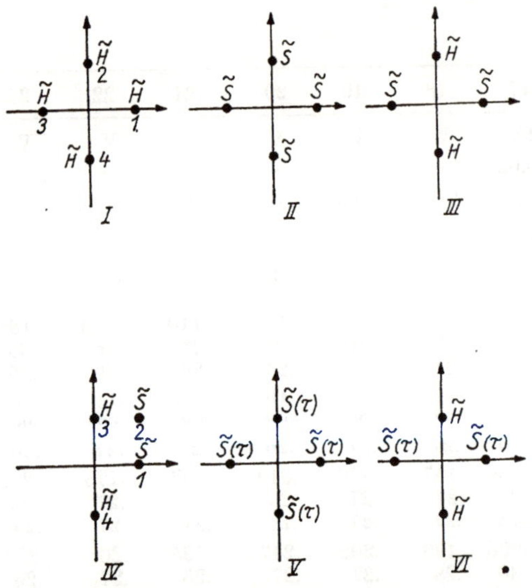

Abb. 2. Typen der Informationsverteilung zur theoretischen Abschätzung (siehe Tab. 2–5).

Tabelle 3
Wirkung des F^S-, K- und τ-Effektes auf numerische Analysen Φ der AT 500 (E-Handanalyse)

F^H	F^s	K_i	τ	$\overline{(E-\Phi)}$	$\overline{E-\Phi}$	$\sqrt{\overline{(E-\Phi)^2}}$	$\overline{(AG-\Phi)}$	$\overline{AG-\Phi}$	$\sqrt{\overline{(AG-\Phi)^2}}$
				30. 4. 75, 00.00					
0,02	keine	SIRS≙	AG	−0,19	3,50	4,95	0	0	0
0,02	0,02	δ_{ij}	0	0,99	2,74	3,67	1,18	1,83	3,15
0,02	0,08	δ_{ij}	0	0,76	2,64	3,68	0,95	1,48	2,57
0,02	0,11	δ_{ij}	0	0,69	2,66	3,73	0,88	1,38	2,39
0,02	0,14	δ_{ij}	0	0,64	2,69	3,78	0,83	1,30	2,25
0,02	0,08	K_1	0	0,61	2,69	3,79	0,80	1,35	2,38
0,02	0,08	K_2	0	0,38	2,99	4,23	4,23	0,57	2,85
0,02	0,08	K_1	0; 4	0,56	2,72	3,84	0,75	1,29	2,30
				1. 5. 75, 00.00					
0,02	keine	SIRS≙	AG	−1,53	3,62	5,60	0	0	0
0,02	0,02	δ_{ij}	0	–	–	–	–	–	–
0,02	0,08	δ_{ij}	0	0,43	2,46	4,14	1,96	2,10	3,83
0,02	0,11	δ_{ij}	0	–	–	–	–	–	–
0,02	0,14	δ_{ij}	0	–	–	–	–	–	–
0,02	0,08	K_1	0	0,29	2,50	4,19	1,82	1,99	3,69
0,02	0,08	K_2	0	–	–	–	–	–	–
0,02	0,08	K_1	0;4	0,27	2,52	4,20	1,80	1,96	3,68

Der Informationswert der SIRS-Daten relativ zur numerischen Analyse, die nur auf aerologischen Geopotentialdaten beruht (AG), kann durch die von den SIRS-Daten ausgelösten Korrekturen abgeschätzt werden. Danach erzeugen die SIRS-Daten bei $F^s=0{,}08$ im Mittel über das ganze Gebiet 1,48 (absolute Abweichung) bzw. 2,57 (RMS) gpdam für den 30. 04. 75 (für den 01. 05. 75 entsprechend 2,10 bzw. 3,83 gpdam) Abweichung. In Einzelpunkten betragen die Abweichungen bis 14 gpdam für den 31. 04. 75 (s. Abb. 3a) und 20 gpdam für den 01. 05. 75. Das unterstreicht die außerordentliche Bedeutung der SIRS-Daten als Zusatzinformation zur aerologischen Information über dem Nordatlantik.

Untersuchen wir nun den Einfluß der Variation der Fehlergröße F_i^s auf das Analyseergebnis. Wie Tab. 3 ausweist, führt eine Erhöhung von $F^s=0{,}02$ auf $F^s=0{,}14$ zu einer Verringerung der von den SIRS-Daten ausgelösten Abweichungen (Korrekturen) bez. der Analyse AG. Die Abweichungen verringern sich für den 30. 04. 75 von 1,83 (abs.) bzw. 3,15 (RMS) gpdam bei $F^s=0{,}02$ auf entsprechend 1,30 bzw. 2,25 gpdam bei $F^s=0{,}14$. In Einzelpunkten entspricht das einer Reduktion der Abweichungen von etwa 3–6 gpdam (für die Werte $F^s=0{,}02$ und $F^s=0{,}08$, siehe Abb. 3a, b). Demnach erreichen die Unterschiede zwischen Analysen, die mit unterschiedlichem F^s berechnet wurden, eine Größe, auf die numerische Prognosemodelle reagieren (s. z. B. [5]). Dabei ist weiter zu bedenken, daß der Fehler einer Einzel-SIRS-Meldung den $F^s=0{,}14$ entsprechenden Wert beträchtlich übersteigen könnte.

Analysiert man die Abhängigkeit der Abweichungen von der Handanalyse, so findet man, daß die geringsten mittleren Abweichungen bei etwa $F^s\approx0{,}08$ erreicht werden. Die mittleren Abweichungsmaße reagieren verständlicherweise recht träge auf Variationen der Fehlergröße F^s.

Schlußfolgernd darf man feststellen, daß die Variabilität der Fehlergröße F^s in korrekter Weise in einem Analyseschema beachtet werden muß.

Der bei der theoretischen Abschätzung erhaltene Befund, daß der Einflußradius einer SIRS-Meldung nicht

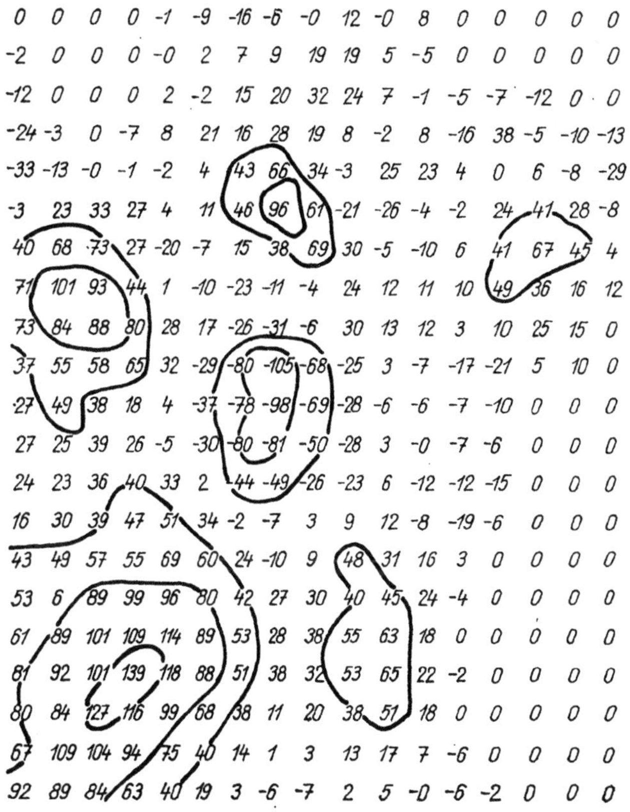

Abb. 3a

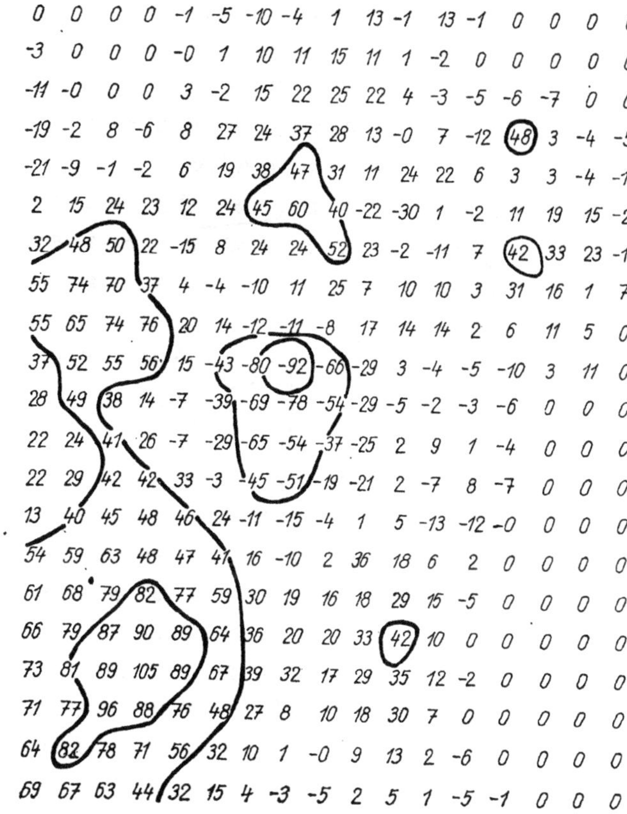

Abb. 3b

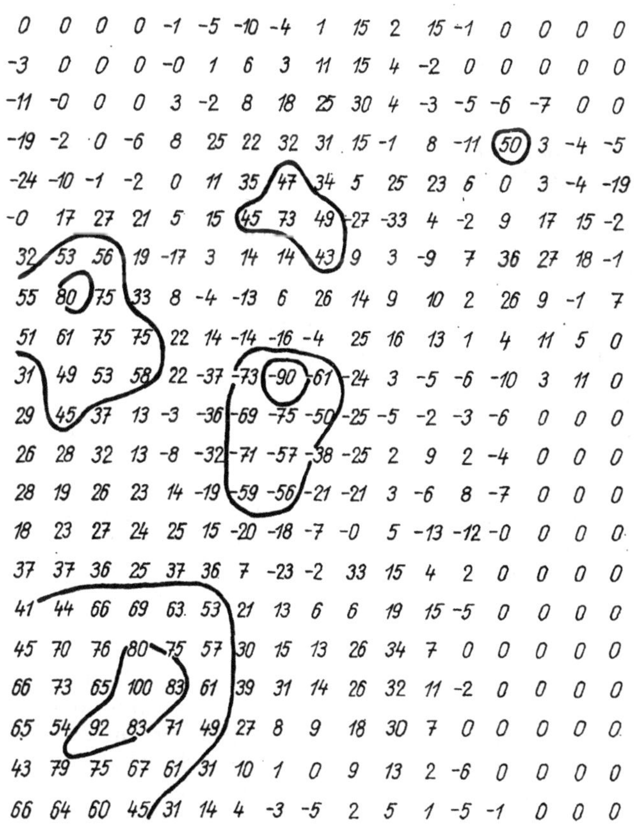

Abb. 3c

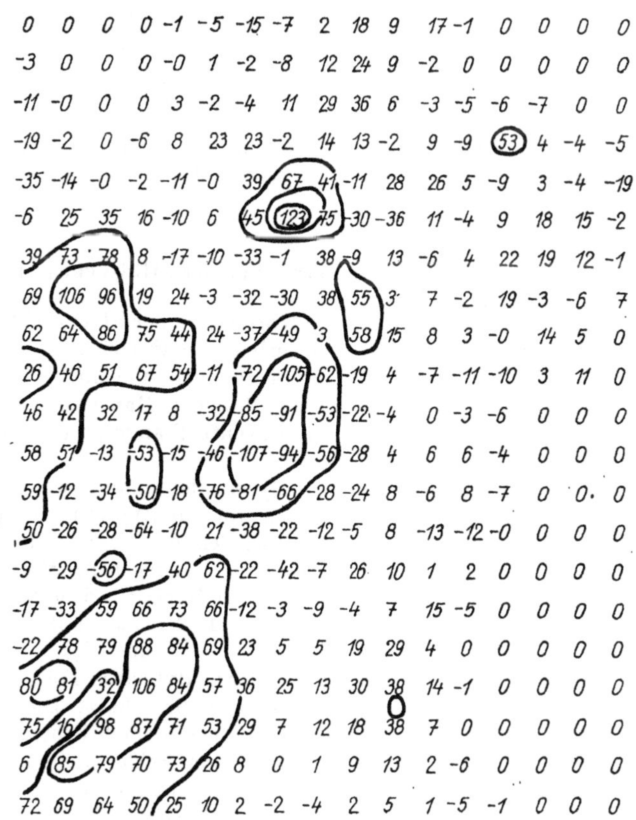

Abb. 3d

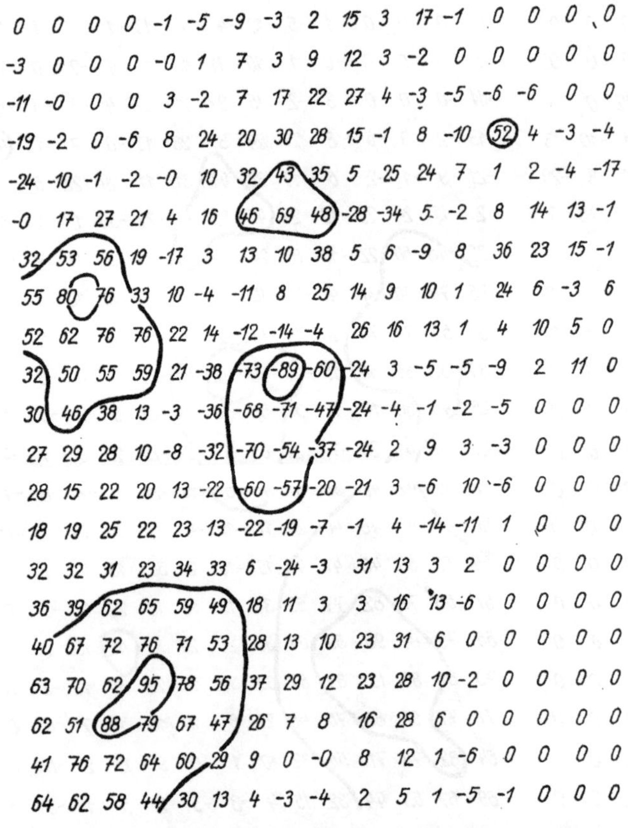

Abb. 3e

Abb. 3 Abweichung der unter Berücksichtigung des F^S-, K- und τ-Effektes berechneten Analysen von AG (Analyse ohne SIRS-Daten allein unter Verwendung der aerologischen Geopotentialdaten).
a – reiner F^s-Effekt, $F^s = 0{,}02$ bei $F^H = 0{,}02$;
b – reiner F^s-Effekt, $F^s = 0{,}08$ bei $F^H = 0{,}02$;
c – kombinierter F^s-Effekt und K-Effekt, $F^s = 0{,}08$, $F^H = 0{,}02$, $K_i = K_1 = e^{-1{,}78\varrho/L}$
d – kombinierter F^s-Effekt und K-Effekt, $F^s = 0{,}08$, $F^H = 0{,}02$, $K_i = K_2 = (1+\varrho/L)e^{-\varrho/L}$
e – kombinierter F^s-, K- und τ-Effekt, $F^s = 0{,}08$. $F^H = 0{,}02$, $K_i = K_1$, $\tau = 0;4$

kreisförmig ist, falls die Analyse gemeinsam mit aerologischen Geopotentialdaten vorgenommen wird, wird bei der Analyse realer Daten bestätigt. So beeinflußt z. B. über See liegende SIRS-Information die Analyse über dem Kontinent nicht wesentlich.

5. Einfluß der Korrelation der SIRS-Fehler (*K*-Effekt)

Die theoretisch zu erwartende Korrelation der SIRS-Fehler wurde erstmals empirisch von *Tarakonova* [13] nachgewiesen. Inzwischen existieren verbesserte Abschätzungen sowohl der horizontalen als auch vertikalen Fehlerkorrelation [3], [9], [12]. Diese Untersuchungen sind noch nicht abgeschlossen. Neuerdings beschäftigen sich auch die SIRS-Datenerzeuger mit diesem Problem. Man muß nämlich annehmen, daß neben einer Abhängigkeit von der Jahreszeit und dem Niveau und eventuell anderen nichttechnischen Faktoren eine solche vom Gerätetyp und der Auswertetechnik besteht.

Entsprechend unserem Analysemodell können wir hier nur die horizontale Korrelation in Betracht ziehen.

Von *Gandin u. M.* [7], [10] wurde zur Approximation der Fehlerkorrelationsfunktion ein Ausdruck der Form $K(\varrho) = e^{-1{,}78\varrho/L}$ verwandt, in [8] werden die Abschätzungen für $K(\varrho) = e^{-\varrho/L}$ durchgeführt (durch einen Schreibfehler wurde in [8] leider $K(\varrho) = e^{-1{,}78\varrho/L}$ angegeben). *Bengtsson* [2] schlug vor, für den Fehler die gleiche Korrelation wie für das Element anzusetzen. Für NIMBUS-6 ergibt sich ein Ausdruck der Form $K_3(\varrho) = e^{-b\,\varrho^2/L^2}$ mit $b \approx 5$, für operative Satelliten wird jedoch $b \approx 1$–2 erwartet (s. [9]). Wir haben demnach Experimente mit $K_1(\varrho) = e^{-1{,}78\varrho/L}$, $K_2 = (1+\varrho/L)\,e^{-\varrho/L}$, $K_3 = e^{-\varrho^2/L^2}$, $K_4 = e^{-5\varrho^2/L^2}$ und $K_5 = e^{-\varrho/L}$ durchgeführt.

Zunächst wurde der Einfluß der analytischen Form der Korrelationsfunktion der SIRS-Fehler auf den Verlauf des theoretischen Interpolationsfehlers abgeschätzt. Die Beachtung der Fehlerkorrelation führt, ganz gleich in welcher Form vorgegeben, generell zu einer beträchtlichen Erhöhung des theoretischen Interpolationsfehlers im Vergleich zur Interpolation mit Daten, deren Fehler nicht korreliert sind (vgl. dazu Tab. 2, Spalte 2 mit Spalten 5–9).

Werden Meldungen, deren Fehler nicht korreliert sind, bei dichtem Beobachtungsnetz optimal kombiniert (interpoliert), dann liegt der Interpolationsfehler bedeutend niedriger als der Beobachtungsfehler der Einzelmeldung (s. Tab. 2, Spalte 2, für $d = 100$ km z. B. und $F_i^H = 0{,}08$ beträgt der Interpolationsfehler 0,02). Dieser Befund gilt für SIRS-Daten nur noch sehr stark abgeschwächt – die Fehlerreduktion ist bedeutend geringer. Ist die Fehlerkorrelationsfunktion mit der des Geopotentials identisch, dann wächst der theoretische Interpolationsfehler mit dem

Abstand d monoton. Fällt sie steiler ab, dann sinkt der Fehler bis zu einer gewissen Entfernung d^*, um danach wieder anzusteigen. Die Größe d^* liegt bei etwa 200 km für K_3, etwa 290 km für K_1 und etwa 350 km für K_4, d. h. mit wachsender Annäherung der Fehlerkorrelationsfunktion an die des Geopotentials verschiebt sich das Minimum gegen Null. Dieses Verhalten gilt es, insbesondere bei einer Feingitteranalyse zu beachten.

Nach den theoretischen Interpolationsfehlern zu urteilen, sind in der Praxis für die etwa gleichwertigen Ansätze K_1, K_3, K_4 einerseits und K_2 andererseits im Mittel Unterschiede von 2–4 gpdam zu erwarten.

Nehmen an der Interpolation nur SIRS-Werte mit dem gleichen Fehler F^s teil, dann reagieren die Interpolationsgewichte auf die K_i nur sehr wenig (s. Tab. 2, Spalten 5–9). Das ändert sich, sobald aerologische und SIRS-Information kombiniert werden (vgl. Tab. 2, Spalte 3 mit Spalten 10–13). Dabei wirkt sich die Form von $K_i(\varrho)$ dort am stärksten aus, wo SIRS-Information dicht benachbart vorliegt (vgl. Tab. 2, Spalten 14–17). Es sei darauf hingewiesen, daß sich die Auswirkungen einer Variation der $K_i(\varrho)$ in dem Maße vergrößern, wie sich der Wert von F_i^s erhöht.

Zur Ergänzung der obigen theoretischen Befunde wurden für beide Datensätze numerische Analysen mit $K_i(\varrho) = K_1$ und für einen Datensatz (30. 04. 75) mit $K_i(\varrho) = K_2$ ausgeführt. In allen Fällen wurde $F_i^s = 0{,}08$ und $F_i^H = 0{,}02$ verwendet.

Wie aus Tab. 3 hervorgeht unterschieden sich die für $K_i = \delta_{ij}$ und $K_i = K_1 = e^{-1{,}78\varrho/L}$ erhaltenen Felder bei $F^s = 0{,}08$ wenig. In Einzelpunkten betragen die entsprechenden Differenzen nur knapp 2 gpdam (vgl. Abb. 3b mit 3c). Diese geringen Unterschiede erklären sich einerseits aus dem relativ kleinen Korrelationsradius von K_1 und andererseits aus den Eigenschaften des SIRS-Informationsangebotes (relativ geringer Umfang, relativ großer mittlerer Meldungsabstand, nur Seegebiete teilweise überdeckt). Auch die Suchtechnik spielt eine Rolle. Wird unter sonst gleichen Bedingungen anstelle von K_1 die Korrelationsfunktion $K_2 = (1 + \varrho/L)\, e^{-\varrho/L}$ verwendet, dann unterscheiden sich sowohl die mittleren Größen (s. Tab. 3) als auch die Einzelwerte (bis zu 6 gpdam) wesentlich (vgl. Abb. 3b mit 3d). Man darf erwarten, daß sich der Fehlerkorrelationseffekt noch stärker ausprägt, wenn F_i^s räumlich variabel angegeben wird. Auf Abb. 3d, rechts unten, wird ein Extrapolationseffekt sichtbar, der durch die Verschlechterung der Bestimmtheit der Korrelationsmatrix beim Übergang von K_1 zu K_2 verursacht wird.

Aus den bisher gewonnenen Ergebnissen darf man folgende Schlußfolgerungen ziehen.

Die Kenntnis der SIRS-Fehler-Korrelationsfunktion ist von wesentlicher Bedeutung. Die SIRS-Datenerzeuger sollten sie den Nutzern mit den Daten selbst zur Verfügung stellen. Dann wäre es möglich, durch entsprechende Festlegung des K_i das Analysesystem den Besonderheiten des Beobachtungssystems anzupassen. Es gäbe aber auch die Möglichkeit, die Auswahl der Prediktoren so zu gestalten, daß auf eine Berücksichtigung der Fehlerkorrelation bei der Interpolation näherungsweise verzichtet werden kann, insbesondere dann, wenn K_i rasch abfällt (s. [8]). Schließlich sei darauf verwiesen, daß die SIRS-Datenerzeuger bei der Festlegung der räumlichen Dichte der an Nutzer abgesetzten Information ebenfalls die Fehlerkorrelation in Betracht ziehen sollten. Obige Betrachtungen erlangen besondere Bedeutung bei einer Feingitteranalyse.

6. Einfluß der Zeitdifferenz (τ-Effekt)

Während das konventionelle Beobachtungssystem streng synoptisch arbeitet, liefern einige in den letzten Jahren an Bedeutung gewinnende Beobachtungsverfahren quasikontinuierlich Information. Ohne auf konkrete technologische Möglichkeiten der Verwertung asynoptischer Information einzugehen, wollen wir hier nur die Frage untersuchen, welchen Informationswert SIRS-Daten besitzen, deren Beobachtungszeit sich um die Zeitspanne τ vom Haupttermin, zu dem die aerologische Information vorliegt, unterscheidet.

Untersuchen wir die Frage zunächst theoretisch mittels der Analyse des theoretischen Interpolationsfehlers für das Informationsangebot V und VI (s. Abb. 2). Um den „reinen" Zeitdifferenzeffekt abzuschätzen, vergleichen wir in Tab. 2 die Spalte 2 mit den Spalten 18–20 und 5–9 mit 21–26. Danach führt eine Zeitdifferenz von $\tau = \pm 3$ Std. zu einer Erhöhung des Interpolationsfehlers für die AT 500 um etwa 1 gpdam, bei $\tau = \pm 6$ Std. um etwa 2 gpdam und bei $\tau = \pm 12$ Std. um 3,5–4,5 gpdam.

Wenn an der Interpolation nur asynoptische Information mit gleichem τ teilnimmt, dann ist der Einfluß von τ und K_i auf die Interpolationsgewichte gering (s. Tab. 2 Spalten 18–26), während sich der Interpolationsfehler beträchtlich ändert. Nimmt neben SIRS-Information auch aerologische an der Interpolation teil, dann reagieren auch die Interpolationsgewichte auf die τ und K_i Variation. Wie aus dem Verlauf der Interpolationsgewichte ersichtlich ist (s. Spalten 27–35 in Tab. 2), nimmt der spezifische Beitrag der asynoptischen SIRS-Information mit wachsendem Meldungsabstand d zu. Dabei erweist sich der Beitrag einer Meldung mit $\tau = \pm\ 12$ Std. als noch sehr informativ. Dennoch sollte man eine derart vom Termin abweichende Information nicht auf die beschriebene Art einbeziehen, sondern die allgemein anerkannte Erfahrung beachten, daß eine dynamische Extrapolation einer statistischen bei Prognoseintervallen > 6 Std. eindeutig überlegen ist. Das Gesagte möchten wir dahingehend zusammenfassen, daß asynoptische Meldungen mit $\tau \approx \pm 3$ Std. als synoptische behandelt werden können, bei $3 < |\tau| \leq 6$ der τ-Effekt in der hier vorgeschlagenen Weise Berücksichtigung finden könnte, während bei $|\tau| > 6$ die Einbeziehung einer Reduktion der Zeitverschiebung durch dynamische Extrapolation mittels eines Prognosemodells unumgänglich erscheint.

Bei den verwendeten Datensätzen beträgt der Maximalwert von τ 4 Std. Vergleicht man die Analysen, bei denen dieser τ-Wert berücksichtigt wurde mit denen, wo $\tau = 0$ gesetzt wurde, dann zeigt Tab. 3 nur geringe Unterschiede in den mittleren Abweichungen. In Einzelpunkten betragen sie weniger als 1 gpdam (s. Abb. 3e). Das bestätigt das oben Gesagte.

7. Zusammenfassende Analyse der Einzeleffekte

Um eine simultane Bewertung der in den Abschnitten 4–6 vorgenommenen isolierten Analyse der Einzeleffekte durchzuführen, schätzen wir einmal die Auswirkung der Effekte auf die erreichbare mittlere Analysegenauigkeit ein (s. Tab. 4) und untersuchen andererseits den spezifischen Beitrag der SIRS-Information, den sie als Zusatzinformation zur aerologischen hervorbringt (s. Tab. 5). Im ersten Fall betrachten wir das Verhältnis der theoretischen Interpolationsfehler, im zweiten Fall das Verhältnis der Interpolationsgewichte. In den Tab. 4 und 5 charakterisieren die Spalte 1 den „reinen" F^s-Effekt, Spalten 2 und 3 den „reinen" K-Effekt bez. F^s-Effekt, Spalten 4 und 5 den kombinierten F^s- und K-Effekt, Spalten 6–8 den „reinen" τ-Effekt bez. F^s-Effekt und Spalten 9–14 den kombinierten F^s-, K- und τ-Effekt.

Tabelle 4
Abschätzung des Einflusses der F^s-, K- und τ-Effekte auf die erreichbare Analysegenauigkeit durch E_i^2/E_j^2 (Verhältnis der in Tab. 2 angegebenen Interpolationsfehler für die Spalten i bzw. j)

Sp. №	1	2	3	4	5	6	7	8	9	10	11	12	13	14
E_i^2/E_j^2	E_2^2/E_1^2	E_5^2/E_2^2	E_8^2/E_2^2	E_5^2/E_1^2	E_8^2/E_1^2	E_{18}^2/E_2^2	E_{19}^2/E_2^2	E_{20}^2/E_2^2	E_{21}^2/E_1^2	E_{22}^2/E_1^2	E_{23}^2/E_1^2	E_{24}^2/E_1^2	E_{25}^2/E_1^2	E_{26}^2/E_1^2
$d=50$	4,0	3,4	3,7	13,4	14,8	1,4	2,6	6,8	15,2	19,8	35,6	16,6	21,2	36,5
200	2,9	2,4	3,4	7,0	9,6	1,3	2,4	5,8	7,8	10,6	20,2	10,5	13,2	22,6
500	1,5	1,3	1,9	1,9	2,8	1,1	1,4	2,7	2,0	2,5	4,3	2,9	3,4	5,1
1000	1,1	1,0	1,2	1,1	1,3	1,0	1,1	1,3	1,2	1,2	1,5	1,4	1,4	1,7

Tabelle 5
Auswirkungen des F^s-, K- und τ-Effektes auf das Gewicht der SIRS-Daten als Zusatzinformation, abgeschätzt durch das Gewichtsverhältnis p_{1i}/p_{1j} (i, j-Spalten № in Tab. 2)

Sp. №	1	2	3	4	5	6	7	8	9	10	11	12	13	14
i/j	3/1	10/3	13/3	10/1	13/1	27/3	28/3	29/3	30/1	31/1	32/1	33/1	34/1	35/1
$d=50$	0,40	0,60	0,60	0,24	0,24	0,80	0,60	0,30	0,20	0,16	0,08	0,20	0,16	0,08
200	0,44	0,73	0,64	0,32	0,28	0,91	0,64	0,27	0,24	0,24	0,12	0,24	0,20	0,10
500	0,70	0,90	0,74	0,63	0,52	0,95	0,79	0,47	0,59	0,52	0,33	0,52	0,44	0,30
1000	0,89	1,00	0,96	0,89	0,86	1,00	0,96	0,84	0,89	0,86	0,75	0,82	0,79	0,71

Aus diesen Angaben lassen sich folgende Schlüsse ziehen. Erstens, jeder der Effekte bewirkt für sich eine Abnahme des Informationswertes der Einzelmeldung bzw. eine Verringerung der erreichbaren mittleren Analysegenauigkeit im Vergleich zum entsprechenden aerologischen Informationsangebot. Der spezifische Beitrag der SIRS-Information als Zusatzinformation verringert sich durch die Wirkung dieser Effekte.

Zweitens, die Auswirkung der Effekte ist zwar der Richtung nach gleich, dem Grade nach aber unterschiedlich. Primäre Bedeutung hat der F^s-Effekt. Der K-Effekt erreicht oder übersteigt dessen Wirkung bei einer Fehlerkorrelationsfunktion mit großem Korrelationsradius. Die Abhängigkeit von der Form der Fehlerkorrelationsfunktion ist ausgeprägt. Der τ-Effekt erreicht erst bei $|\tau| > 6$ Std. die Größe der beiden Vorgenannten.

Drittens, die Wirkung der Effekte hängt von der Meldungsdichte (charakterisiert durch die Größe d) ab. Je größer d ist, um so geringer die Auswirkung der Effekte auf die Analysegenauigkeit und auf die Verringerung des spezifischen Beitrages als Zusatzinformation. Tab. 5 demonstriert außerdem anschaulich, daß SIRS-Information als Zusatzinformation erst bei geringer Netzdichte der aerologischen Sondierungen bedeutendes Gewicht erhält.

Wie frühere Untersuchungen [8] zeigen, dürften die hier gezogenen Schlußfolgerungen auch für ein komplexeres Informationsangebot gültig bleiben.

Bei der praktischen numerischen Analyse sollte die Beachtung des F^s- und K-Effektes, so lehren die erhaltenen Ergebnisse, im Vordergrund stehen, solange $|\tau| \leq 3$ Std. beträgt. Deshalb ist es notwendig, das sei noch einmal betont, daß die SIRS-Datenerzeuger die entsprechenden Fehlercharakteristiken dem Nutzer zur Verfügung stellen.

Literatur

[1] *Bagrov, A. N., S. L. Belousov, A. G. Tarnopol'skij*, Schema ob-ektivnogo analiza polej davlenija, baričeskoj tendencii, temperatury i točki rosy po dannym seti sinoptičeskich stancij. Trudy GMC, vyp. 91, 1972, 58–75.

[2] *Bengtsson, L.*, 4-Dimensional Assimilation of Meteorological observations. GARP-Publ. Ser. № 15, 1975, 76 pp.

[3] *Bengtsson, L.*, Use of Satellite Data in Numerical Models (Presented at Royal Met. Society 19. 5. 1976) 47 pp.

[4] *Bengtsson, L., P. Morel*, The Performance of Space Observing Systems for the First GARP Global Experiment. The GARP Programme on Numerical Experimentation, Report № 6, 1974, 58 pp.

[5] *Fuks-Rabinovič, M. S.*, Ob ocenke vlijanija točnosti ischodnoj informacii na prognoz geopotentiala i skorosti vetra s ispol'zovaniem primitivnoj barotropnoj modeli atmosfery. Trudy GMC SSSR, vyp. 71, 1970, 98–108.

[6] *Gandin, L. S., R. L. Kagan*, O postroenii sistemy ob-ektivnogo analiza raznorodnych dannych na osnove optimal'noj interpoljacii i optimal'nogo soglasovanija. Meteor. i Gidrol. № 5, 1974, 3–10.

[7] *Gandin, L. S., K. M. Lugina*, Ob učete dannych kosvennogo zondirovanija atmosfery pri četyrechmernom analize meteorologičeskych polej. Trudy GGO, vyp. 348, 1975, 3–15.

[8] *Kluge, J.*, Optimal combination of different kinds of geopotential information by a one-step and multi-step procedure. Proceedings of the JOC Study Group Conference on Four-Dimensional Data Assimilation, Nov. 1975. The GARP-Programme on Numerical Experimentation, Report № 11, 1976. 241–249.

[9] Report of the Thirteenth Session of the GARP Working Group on Numerical Experimentation, Geneva May 1976, 19 pp.

[10] *Šachmejster, V. A.*, Čislennye ėksperimenty po sovmestnomu učetu aėrologičeskoj, sputnikovoj i prognostičeskoj informacii. Trudy GGO vyp. 348, 1975, 16–30.

[11] *Skoda, M.*, Prisperek k Metodam Objectivni Analyzy Meteorologickych Udaju. Hydrometeorologicky Ústav, Praha 1975, 268 S.

[12] *Tarakonova, V. P.*, O zakonomernostjach ošibok opredelenija temperatury i geopotentiala po dannym kosvennogo zondirovanija atmosfery. Trudy GGO, vyp. 348, 1975, 46–52.

[13] *Tarakonova, V. P.*, K voprosu o točnosti temperaturnogo zondirovanija atmosfery s iskustvennych sputnikov zemli. Meteor. i Gidrol. № 4, 1974, 76–78.

Anschrift: Dr. J. Kluge und Dipl.-Met. M. Petzold, Meteorologischer Dienst der DDR, Zentrale Wetterdienststelle, DDR – 15 Potsdam, Michendorfer Chaussee 23

551.508.71

Feuchtigkeitsmessung mit Bariumfluorid-Meßfühlern

I. Aufbau und Eigenschaften eines Miniatur-Bariumfluorid-Feuchtigkeitsmeßfühlers

Von **Th. Foken, N. Hippmann, U. Lehmann** und **D. Sonntag**

Mit 5 Abbildungen

Zusammenfassung: Es wird ein Adsorptions-Feuchtigkeitsmeßelement mit Bariumfluorid als feuchtigkeitsaktiver Substanz beschrieben. Die Größe beträgt 3,0 mm × 4,5 mm. Der Widerstand bei 100% relativer Feuchtigkeit und 30 °C beträgt ca. 10 kΩ und die Empfindlichkeit 25% relative Feuchtigkeit pro Dekade Widerstandszunahme. Die temperaturabhängige Zeitkonstante beträgt bei 10 . . . 30 °C ca. 1 . . . 0,1 s.

Summary: This paper is concerned with the description of an adsorption humidity measuring element with bariumfluoride as the humidity sensitive substance, its size being 3.0 to 4.5 mm. The resistance at a relative humidity of 100 per cent and 30 °C is about 10 kΩ and the sensitivity is 25 per cent relative humidity per decade of increase of resistance. The temperature-dependent time constant is about 1 to 0.1 sec between 10 . . . 30 °C.

Резюме: Описан датчик для измерения влажности, который работает на принципе адсорбции с фторидом бария. Размеры датчика 3,0 mm × 4,5 mm. Сопротивление при относительной влажности 100% и температуре 30° Ц. около 10 kΩ и чувствительност составляет 25% относительной влажности на один порядок величны увеличения сопротивления. Постоянная времени, зависящая от температуры, составляет при температурах 10 . . . 30 °Ц. приблизительно 1 . . . 0,1 сек.

1. Einleitung

Feinstrukturuntersuchungen des Windes, der Temperatur und der Luftfeuchtigkeit in der wassernahen Luftschicht sind von ausschlaggebender Bedeutung für die weitere Verbesserung der Methoden zur Bestimmung des Energieaustausches zwischen Meer und Atmosphäre. Dabei sind Untersuchungen zum Mechanismus des Energieaustausches in den untersten Zentimetern über dem bewegten Wasser sehr wichtig. Meßtechnisch konnte dieses Problem bislang nur für die Temperatur (s. *Th. Foken* 1975) gelöst werden. Für die Feuchtigkeitsmessung kann das Psychrometerprinzip von *A. H. Schooley* 1975 nicht befriedigen, bei dem ein Thermistor beim Eintauchen in das Wasser befeuchtet wird und beim Wiederauftauchen bis zur Abtrocknung die Feuchttemperatur mißt. Bis zum Eintauchen wird dann wieder die Lufttemperatur gemessen. Die sehr empfindlichen UV- und IR-Hygrometer sind zu groß, um für derartige Messungen eingesetzt werden zu können. Aus diesen Gründen wurde nach entsprechenden Literaturrecherchen in der Arbeitsgruppe Ozeanologie der Sektion Physik der Karl-Marx-Universität Leipzig mit der Entwicklung eines Miniatur-Adsorptionshygrometers mit Bariumfluorid als feuchtigkeitsaktiver Substanz begonnen. Dabei konnte sich auf Vorarbeiten, allerdings für wesentlich größere Fühler, von *F. E. Jones* und *A. Wexler* 1960, *F. E. Jones* 1962 und *D. Sonntag* 1968, 1970 gestützt werden.

2. Aufbau des Bariumfluorid-Feuchtigkeitsfühlers

2. 1. Anwendbarkeit von Bariumfluorid zur Feuchtigkeitsmessung

Die Verwendung von Ad- und Absorptionshygrometern für die Feuchtigkeitsmessung ist weit verbreitet (s. *D. Sonntag* 1966–1968). Dabei finden besonders Lithiumchloridfühler Anwendung, z. B. in der Radiosondentechnik. Eine genaue Analyse aller möglichen Einsatzsubstanzen (s. *F. E. Jones* 1962, *D. Sonntag* 1970) zeigt allerdings, daß sich hinsichtlich nachfolgender Kriterien Bariumfluorid am besten eignet, da es diese weitestgehend erfüllt:

- Eichfähige Meßelemente,
- Genauigkeit ±5% relative Feuchtigkeit,
- Meßwert als elektrische Größe,
- Hystereseeffekte innerhalb der Meßgenauigkeit,
- annähernde Temperaturunabhängigkeit,
- wasserunlösliche Substanz,
- geringe Größe und geringe Masse,
- geringe Kosten.

Bariumfluorid ist ein Schwermetallsalz, dessen Ionenbindung einen stark heteropolaren Charakter hat. Die elektrische Leitfähigkeit wird durch das Vermögen zur Hydratation hervorgerufen, womit gleichzeitig die Feuchtigkeitsabhängigkeit des elektrischen Widerstandes gegeben ist. Da die Hydratation im wesentlichen nur durch Adsorption an der Grenzfläche zur Umgebung auftritt, reagiert Bariumfluorid auf Feuchtigkeitsänderungen schnell und nur mit geringen Hystereseerscheinungen. Begünstigt wird dies durch eine sehr geringe Bedampfungsdicke. Sehr feine Fühlerstrukturen führen allerdings zu einer hohen Individualität bei gleicher Bedampfung wegen der unterschiedlichen Störstellenanzahl in den Kristallen auf Grund des Nichterreichens der statistischen Gleichverteilung. Die geringe Wasserlöslichkeit bedingt eine gute Haltbarkeit der Oberflächenbedampfung.

2. 2. Aufbau des vorgeschlagenen Feuchtigkeitsfühlers

Auf Grund der beschriebenen Einsatzbedingungen in der wassernahen Luftschicht eignet sich der relativ große Fühler nach *D. Sonntag* 1970 nicht für eine Anwendung. Es sollte daher ein möglichst kleiner Fühler, mit ähnlichen Charakteristiken, wie sie der große Fühler hat, entwickelt werden:

- Breite des wirksamen Mäanders $< 0{,}1$ mm,
- Zeitkonstante < 1 s,
- Widerstand bei 100% relativer Feuchtigkeit $10^2 \ldots 10^4\ \Omega$,
- große Empfindlichkeit
- geringe Temperaturabhängigkeit.

Auf Grund dieser Forderungen wurden als Maße für den Fühler festgelegt:

- Fühlergröße 3,0 mm × 4,5 mm,
- Mäanderbreite 50 µm,
- wirksame Elektrodenfläche $\approx 10^{-3}$ mm².

Als Trägermaterialien eignen sich JENAer RASOTHERM-Glas, JENAer Geräteglas 20 und Quarzglas, da diese Gläser weitgehend alkalifrei sind. Die Größe des verwendeten Glases beträgt 49 mm × 49 mm × 2 mm. Dabei wurden auf ein derartiges Glassubstrat 48 Fühler aufgedampft. Die Abb. 1 zeigt einen Fühler im Vergleich zu dem großen Messfühler, und in Abb. 2 ist eine Teilvergrößerung des Chrommäanders zu sehen.

Abb. 1. Größenvergleich des Miniatur-Bariumfluorid-Feuchtigkeitsfühlers (links in der Schachtel) mit dem Fühler nach *D. Sonntag* 1970

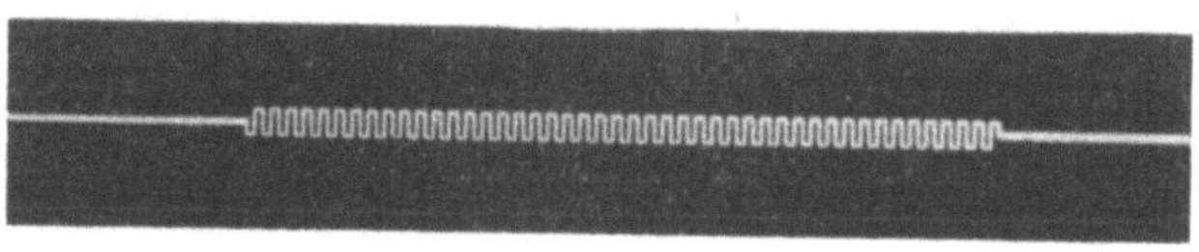

Abb. 2. Ausschnittsvergrößerung des Chrommäanders (Vergrößerung 33 : 1)

Nach einer fotochemischen Abdeckung der Trennlinie wurde die Chrombedampfung durchgeführt. Für die Bariumfluoridbedampfung wurden die Bedingungen im Hinblick auf optimale Fühlereigenschaften variiert, wobei die Bedampfungsdicke im Bereich 0,1 . . . 1 µm festgelegt wurde. Die Aufbewahrung der Fühler erfolgt in trockener Luft über Silikagel.

3. Eigenschaften der Fühler

3. 1. Meßschaltung und Eichung

In der Meßschaltung für die Eichung ist der Bariumfluorid-Feuchtigkeitsfühler als verstärkungsbestimmender Widerstand in einen invertierenden Verstärker geschaltet, der mit einer konstanten Wechselspannung zur Vermeidung von Polarisationserscheinungen am Fühler betrieben wird (Abb. 3). Die Verstärkung dieser eingeprägten Spannung erfolgt entsprechend der Beziehung

$$V = \frac{R_G}{R} = \frac{U_a}{U_e}.$$

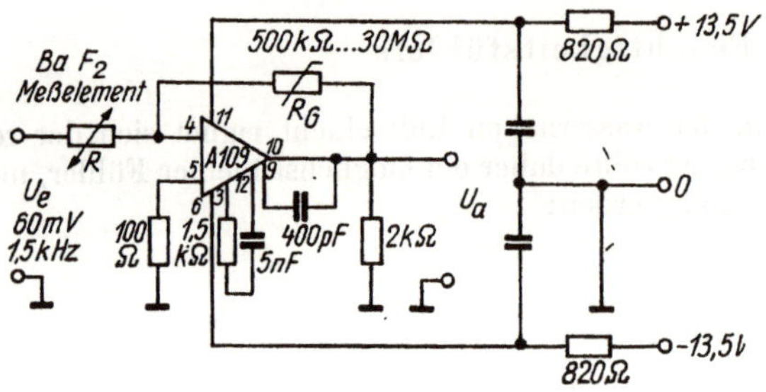

Abb. 3. Meßschaltung

Die Messung der Ausgangsspannung erfolgt mit einem Röhrenvoltmeter. Um ständig in einem Bereich optimaler Verstärkung arbeiten zu können, ist der Gegenkopplungswiderstand verstellbar. Wegen des Eingangswiderstandes des Operationsverstärkers von 0,5 MΩ muß magn $R \leqq 0{,}5$ MΩ werden, um merkliche Anpassungsfehler auszuschließen. Die Eichung erfolgte bei 10 °C, 20 °C und 30 °C im Bereich 50 . . . 100% relative Feuchtigkeit mit zunehmender und anschließend abnehmender Feuchtigkeit, um Hystereseeffekte feststellen zu können. Dieser Eichbereich entspricht den Anwendungsbedingungen. Die entsprechenden Eichkurven eines Fühlers sind in Abb. 4 dargestellt.

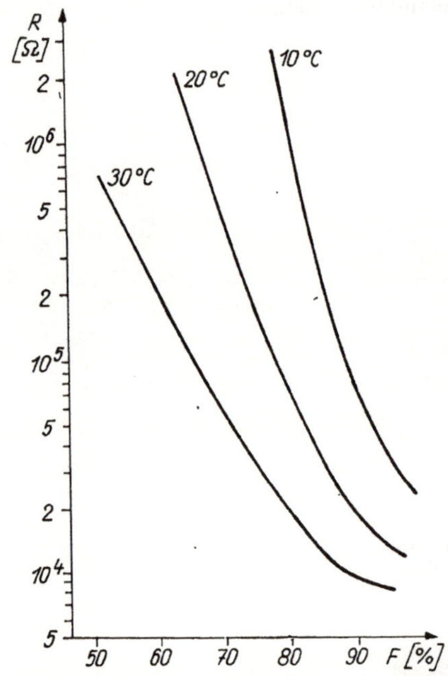

Abb. 4. Temperaturabhängigkeit der Eichkurven

3. 2. Eigenschaften

Die wichtigsten Kriterien zur Charakterisierung der Fühler sind:
- Widerstand bei 100% relativer Feuchtigkeit und 30 °C,
- Änderung der relativen Feuchtigkeit pro Dekade Widerstandsänderung,
- Temperaturabhängigkeit,
- Zeitkonstante,
- Hysterese,
- Log-Linearität,
- Alterungserscheinungen.

Die bisherigen Untersuchungen dienten insbesondere der Festlegung einer optimalen Bariumfluoridbedampfung. Dabei konnte festgestellt werden, daß der Widerstand bei 100% relativer Feuchtigkeit mit zunehmender Bedampfungsdicke abnimmt. Demgegenüber existiert aber eine Bedampfungsstärke mit einer maximalen Empfindlichkeit von ca. 25% pro Dekade Widerstandsänderung. Der entsprechende Widerstand beträgt $R \approx 10$ kΩ.

Von außerordentlicher Wichtigkeit ist die Untersuchung über den Temperatureinfluß. Wegen der Meßaufgaben sollen Temperaturänderungen einen merklich kleineren Einfluß auf die Widerstandsänderungen haben als Feuchtigkeitsänderungen. Die Abhängigkeit lautet

$$d \lg R = \left(\frac{\partial \lg R}{\partial F}\right) dF + \left(\frac{\partial \lg R}{\partial T}\right) dT,$$

wobei als mittlere Beziehung bestimmt wurde

$$\Delta \lg R = \frac{-0{,}03}{1\%} \Delta F + \frac{-0{,}03}{1K} \Delta T .$$

Auf Grund dieser Ergebnisse bedarf es jedoch der ständigen Prüfung der Größenordnung der Temperatur- und Feuchtigkeitsänderungen, um abschätzen zu können, ob bei den Messungen Korrekturen notwendig werden. Eine parallele Temperaturregistrierung ist zweckmäßig.

Wie die Eichkurve in Abb. 4 zeigt, ist die Log-Linearität weitgehend gewährleistet. Wie schon bei *D. Sonntag* 1970 ersichtlich, treten merkliche Abweichungen von der Log-Linearität insbesondere nahe 100% relativer Feuchtigkeit auf. Die Log-Linearität sollte daher nur für den Bereich $F < 95\%$ annähernd angenommen werden. Eine ebenfalls schwer erfaßbare Größe ist die Hysterese, die jedoch relativ gering ist. Beide Effekte stellen keine merkliche Beeinträchtigung der Anwendbarkeit dar.

Alterungserscheinungen konnten nur in einem sehr geringen Umfang untersucht werden. Es zeichnet sich jedoch schon klar eine Widerstands- und Empfindlichkeitszunahme mit der Zeit ab. Diese können durch Tempern bei 400 °C wieder rückgängig gemacht werden (Abb. 5).

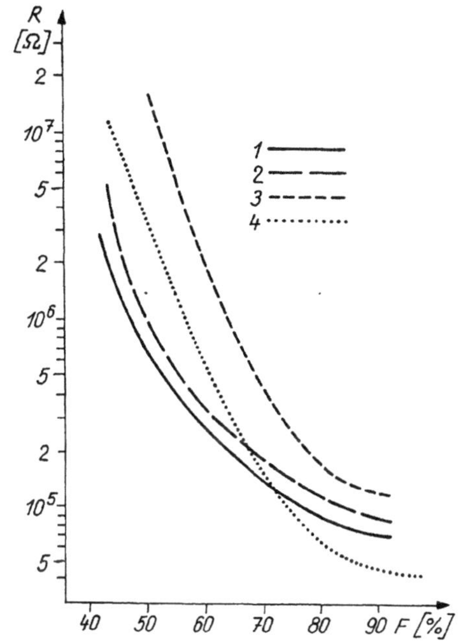

Abb. 5. Alterungsabhängigkeit der Eichkurven
1 nach der Bedampfung
2 nach einer Woche
3 nach 5 Monaten
4 nach 5 Monaten und 10 Minuten Tempern bei 400 °C

Die starke Temperaturabhängigkeit der Zeitkonstante, wie sie *F. E. Jones* 1967 beschrieb, gilt für die Miniaturfühler sogar noch in einem stärkeren Maße. So beträgt die Zeitkonstante bei 30 °C 0,1 s, während sie bei 10 °C nur noch ca. 1 s ist. Die geringsten Zeitkonstanten werden bei kleinen Bedampfungsdicken festgestellt, jedoch überwiegt dann der Temperatureffekt.

Die in der Tabelle 1 angegebenen Eigenschaften entsprechen weitgehend denen des Fühlers nach *D. Sonntag* 1970. Die Miniaturisierung brachte allerdings keine Senkung des Fühlerwiderstandes. Offensichtlich existiert ein Optimum für den Fühlerwiderstand, bei dem die Log-Linearität am besten erfüllt ist.

Tabelle 1

Eigenschaften der Miniatur – Bariumfluoridfühler

Eigenschaften	Wert
Widerstand bei 100% relativer Feuchtigkeit und 30 °C	≈10 kΩ
Empfindlichkeit	
Prozent relativer Feuchtigkeit pro Dekade Wid.-Änderung	20 . . . 30%
$\Delta \lg R/1\%$	−0,03
$\Delta \lg R/1$ K	−0,03
Zeitkonstante	
30 °C	0,1 s
10 °C	1 s
Hysterese	gering
Alterung	Widerstands- und Empfindlichkeitszunahme mit der Zeit, Herstellung des Ausgangszustandes durch Tempern bei 400 °C

4. Anwendung

Die Abhängigkeit des Widerstandes von der Feuchtigkeit ist exponentiell. Es ist daher für die praktische Anwendung ein spezielles Meßgerät zu konstruieren (möglichst mit logarithmischem Verstärker), um eine lineare Abhängigkeit des Ausgangssignals von der relativen Feuchtigkeit zu erhalten. Durch den hohen Fühlerwiderstand ist der elektronische Aufwand jedoch erheblich.

Die Anwendung der Miniatur-Bariumfluorid-Feuchtigkeitsfühlers ist aber keinesfalls auf die angegebene Feuchtigkeits-Mikrostrukturuntersuchung der wassernahen Lustschicht beschränkt. Er läßt sich vielmehr bei allen Relativmessungen und bei Absolutmessungen von geringer Dauer anwenden. Wesentliche Vorteile sind die geringen Abmessungen des Fühlers und die relativ einfache Handhabung. Anwendungsbereiche sind daher neben der Meteorologie (speziell Radiosondentechnik und angewandte Richtungen wie Agrarmeteorologie und Technische Meteorologie) auch verschiedene Bereiche der Industrie, Medizin, Biologie, usw.

Die weiteren Untersuchungen sollen insbesondere der Eichstabilität und verbesserten elektronischen Schaltungen dienen sowie der Anwendung für verschiedene Meßaufgaben. Die Ergebnisse sollen in dieser Zeitschrift veröffentlicht werden (*Th. Foken 1978*).

5. Danksagung

Die Autoren danken Herrn Doz. Dr. sc. *P. Hupfer* für die allseitige Förderung und Unterstützung der Entwicklungsarbeiten, Herrn Doz. Dr. *G. Otto* für die Ermöglichung der Bedampfungsversuche und den Herren Ing. *G. Neubert* und Feinmechanikermeister *W. Bohmann* für die wertvollen praktischen Hinweise und Unterstützungen. Weiterhin gilt der Dank dem VEB Röhrenwerk „Anna Seghers", Neuhaus am Rennweg und dem VEB Carl-Zeiss-Jena für die Unterstützung bei der Fühlerherstellung.

Literatur

Foken, Th.: Die Messung der Mikrostruktur der vertikalen Lufttemperaturverteilung in unmittelbarer Nähe der Grenze zwischen Meer und Atmosphäre. Z. Meteor., Berlin *25* (1975) 5, 292–295.

Foken, Th.: Feuchtigkeitsmessung mit Bariumfluorid-Meßfühlern, II Aufbau. Eigenschaften und Einsatz eines Meßgerätes mit Miniatur-Bariumfluorid-Feuchtigkeitsmeßfühler. Z. Meteor., Berlin *28* (1978) im Druck.

Jones, F. E.: Evaporated-Film Electric Hygrometer Elements. J. of Research of the National Bureau of Standards – C. Engineering and Instrumentation *66 C* (1962) 3, 209–216.

Jones, F. E.: Study of the Storage Stability of the Barium Fluoride Film Electric Hygrometer Element. J. of Research of the National Bureau of Standards – C. Engineering and Instrumentation *71 C* (1967) 3. 199–207.

Jones, F. E. and *A. Wexler,:* A Barium Fluoride Film Hygrometer Element. J. Geophys. Research, Boston *65* (1960) 7, 2087–2095.

Schooley, A. H.: Simple Method for Measuring Relative Humidity, Water and Air Temperature Within a Few Millimeters of Wind-Generated Water-Waves. J. Phys. Ocean., Boston 5 (1975) 3, 519–522.

Sonntag, D.: Hygrometrie – Ein Handbuch der Feuchtigkeitsmessung in Luft und anderen Gasen. Berlin: Akademie-Verlag 1966–1968, 1086 S.

Sonntag, D.: Verfahren zur Herstellung eines Elementes für die elektrische Bestimmung der relativen Luftfeuchtigkeit. DDR-Wirtschaftspatent Nr. 64909 vom 5. 12. 1968.

Sonntag, D.: Ergebnisse über die Herstellung und einjährige Erprobung von Bariumfluorid-Feuchtigkeitsmeßelementen. Z. Meteor., Berlin *21* (1970) 8, 231–237.

Anschriften: Dipl.-Met. Thomas Foken, Dipl.-Phys. Norbert Hippmann, Karl-Marx-Universität Leipzig, Sektion Physik, Arbeitsgruppe Ozeanologie, DDR–701 Leipzig, Talstraße 35.
Uta Lehmann, Karl-Marx-Universität Leipzig, Sektion Physik, Arbeitsgruppe Angewandte Kernphysik, DDR-701 Leipzig, Linnéstraße 5.
Dr. habil. Dietrich Sonntag, Meteorologischer Dienst der DDR, Instrumentenamt, DDR–15 Potsdam, Michendorfer Chaussee 23.

551.593.54(430.2-37.7-21)
551.521.3
551.588.7

Messungen der Extinktion der Sonnenstrahlung in der atmosphärischen Bodenschicht des Stadtgebietes Leipzig

Von **Ch. Hänsel**

Mit 2 Abbildungen

Zusammenfassung: Im Stadtzentrum gemessene Trübungsfaktoren liegen in der Heizperiode um ca. 20% („Gesamttrübungsfaktor") bzw. 40% („Kurztrübungsfaktor") über den Trübungsfaktoren der heizfreien Zeit. Während der Heizperiode werden sehr hohe Werte des „speziellen Trübungsfaktors" der atm. Bodenschicht durch gleichzeitige Aktinometermessungen an der Erdoberfläche und in 120 m Höhe gefunden, wobei die relativ hohen Werte des speziellen Kurztrübungsfaktors auf die starke Extinktion durch Partikeln verweisen. Außerdem wird versucht, wellenlängenspezifische Extinktionskoeffizienten der atm. Bodenschicht mit einem Quarzspektrographen zu messen.

Summary: In the centre of the town the turbidity factors during time of heating are higher by ca. 20% („general turbidity factor") and 40% („shortwave turbidity factor") than the turbidity factors during summer time. During the heating time very high values of the „special turbidity factor" of the boundary layer are found by simultaneous actinometric measurements at the ground and at an altitude of 120 m, whereby the relatively high values of the „special shortwave turbidity factor" indicate the strong extinction by particles. Moreover, the attempt is made to measure the extinction coefficient, which is dependent on the wavelength, in the atmospheric boundary layer by a quartz spectrograph.

Stadt- und Industriegebiete sind wegen der Aerosolanreicherung in der atmosphärischen Grundschicht zu einem Experimentierfeld für Forschungsarbeiten zum Strahlungs- und Wärmehaushalt und damit zur Modellierung klimabildender bzw. -verändernder Prozesse geworden. Die zahlreichen Beiträge zum Problem „Stadtklima" in der Fachliteratur liefern vielfältige Angaben über stadttypische Anomalien. Sie weisen unter anderem Verminderungen der Globalstrahlung durch die erhöhte Trübung einer Stadtatmosphäre gegenüber der unbeeinflußten Umgebung von durchschnittlich 10 . . . 15% aus (WMO, 1970).

Im Stadtgebiet Leipzig, das durch die Konzentration von Braunkohlentagebau und Nachfolgeindustrie im umgebenden Territorium mit Luftverunreinigungen besonders stark belastet ist, wurden in der Zeit von Juni 1976 bis Februar 1977 aus Aktinometermessungen im Stadtzentrum an 26 Tagen (mit insgesamt 153 Einzelmessungen) die Trübungsfaktoren (nach *Linke*) bestimmt. In der Tabelle 1 sind die Mittelwerte des Gesamttrübungsfaktors $\bar{T}_g$ und des Kurztrübungsfaktors $\bar{T}_k$ angegeben, dazu die absoluten Extremwerte aus der Stichprobe. In üblicher Weise bezieht sich der Kurztrübungsfaktor auf den kurzwelligen Anteil des Sonnenspektrums bis zur Wellenlänge 620 nm. Eine Unterteilung in Meßwerte aus der Heizperiode und der heizfreien Periode zeigt, daß die erhöhte Verschmutzung der atmosphärischen Grundschicht in der Heizperiode zu einem um ca. 20% ($\bar{T}_g$) bzw. 40% ($\bar{T}_k$) gegenüber der heizfreien Zeit höheren Trübungsfaktor führt. (Die Mittelwerte der Trübungsfaktoren in der Heizperiode und in der heizfreien Zeit unterscheiden sich statistisch signifikant bei einer Irrtumswahrscheinlichkeit von 10% bei dem Gesamttrübungsfaktor und von 3% bei dem Kurztrübungsfaktor). Vergleichsweise ergeben die Trübungsmessungen in Potsdam aus den Jahren 1961/70 einen Gesamtmittelwert $\bar{T}_g = 4{,}22$ mit einem Mittelwert in der Heizperiode von 3,82 und in der heizfreien Zeit von 4,79. Unter dem Stadteinfluß verhalten sich die Trübungsfaktoren somit gegenläufig zum normalen Jahresgang (s. *Foitzik* und *Hinzpeter*, 1958).

Im Zeitabschnitt zwischen Oktober 1976 und Februar 1977, also nur innerhalb der Heizperiode, wurden durch synchrone Aktinometermessungen am Fuße und auf dem Dach des Hochhauses der Karl-Marx-Universität, unmittelbar im Stadtzentrum, bei einem Höhenunterschied der Meßpunkte von 120 m, an 10 Tagen mit insgesamt 33 Messungen die „speziellen Trübungsfaktoren" der zwischenliegenden atmosphärischen Bodenschicht bestimmt (*Linke*, 1929). Vergleichbar dem üblichen Trübungsfaktor für die gesamte Atmosphäre, stellt der spezielle Trübungsfaktor das Vielfache der Trübung einer Schicht dar in bezug auf die Trübung, die in sauberer Luft in der Schicht (Rayleigh-Atmosphäre) vorhanden wäre. Für relativ dünne Schichten gilt die bereits von *Lauscher* (1930) angegebene Bestimmungsgleichung

$$\tau = T_1 + (T_2 - T_1) \frac{p_2}{p_2 - p_1},$$

wenn T_1, p_1 Trübungsfaktor und Luftdruck an der Obergrenze und T_2, p_2 an der Untergrenze der Schicht bedeuten.

Aus der Stichprobe ergeben sich die ebenfalls in der Tabelle 1 enthaltenen Mittel- und Extremwerte des speziellen Gesamttrübungsfaktors (τ_g) und des speziellen Kurztrübungsfaktors (τ_k). Während bei dem allgemeinen Trübungsfaktor in der Heizperiode das Verhältnis von Kurztrübungsfaktor zu Gesamttrübungsfaktor 0,60 beträgt, steigt es beim speziellen Trübungsfaktor auf 0,76 an. Die sowohl absolut als auch relativ sehr hohen Werte der speziellen Kurztrübungsfaktoren beweisen, daß in der Bodenschicht vornehmlich die Partikelextinktion wirkt.

Der relative Anteil der Trübungswirkung der 120 m dicken atmosphärischen Bodenschicht an der gesamten Trübung der Atmosphäre (also $(T_2 - T_1)/T_2$) ergibt im Mittel der Stichprobe für die Gesamttrübung 0,19, für die Kurztrübung 0,22, wobei diese Werte zwischen den absoluten Extremwerten 0,08 und 0,45, bzw. 0,08 und 0,39 liegen. Bei allen Messungen betrug die relative Luftfeuchte weniger als 80%.

Tabelle 1

Mittlere Trübungsfaktoren und mittlere spezielle Trübungsfaktoren der atm. Bodenschicht, Leipzig, Juni 1976–Februar 1977

	heizfreie Periode		Heizperiode		ges. Jahr
T_g	4,89		5,80		5,41
max, min	8,05	3,26	9,82	3,46	
T_k	2,51		3,49		3,08
max, min	4,68	1,41	6,14	1,85	
$\bar{\tau}_g$			81,8		
max, min			184,9	28,1	
$\bar{\tau}_k$			62,0		
max, min			135,0	18,53	

Eine Korrelationsanalyse zwischen dem speziellen Trübungsfaktor und den wesentlichen meteorologischen Einflußgrößen Windgeschwindigkeit und vertikaler Temperaturgradient in der atmosphärischen Bodenschicht liefert für die Stichprobe (negative) partielle Korrelationskoeffizienten, die dicht an der Signifikanzgrenze bei 5% Irrtumswahrscheinlichkeit liegen, wobei die Abhängigkeit der Trübung von der Schichtungsstabilität stärker hervortritt als die Abhängigkeit von der Windgeschwindigkeit. Aus einer umfangreicheren Stichprobe sind straffere und statistisch gut gesicherte Korrelationen zu erwarten.

Um die Extinktion der atmosphärischen Bodenschicht auch in einzelnen Wellenlängenbereichen zu erfassen, wurden Spektren der Sonnenstrahlung am Fuße und auf dem Dach des Hochhauses mit einem Quarzspektrographen nach *Andrejtschin* und *Kechlibarov* (1965) aufgenommen. Das Experiment diente in erster Linie zur Erprobung der Brauchbarkeit dieser Methode. Der Quarzspektrograph ist mit einer Kleinbildkamera kombiniert. Das auf einem Negativfilm festgehaltene Spektrum eignet sich wegen der hierfür zu ungenau definierten spektralen Filmempfindlichkeit nicht für absolute Intensitätsmessungen, liefert aber bei sorgfältiger photometrischer Auswertung brauchbare relative Intensitätswerte. Dazu sind natürlich übereinstimmende Belichtungszeiten und strenge Einhaltung definierter Entwicklungsbedingungen des Films vorausgesetzt.

Aus einer kleinen Meßserie, die im Sommer 1975 durchgeführt wurde, ließen sich nach der Beziehung

$$a(\lambda) = \ln(I_1/I_2)/h$$

(I_1, I_2 = Intensitätsverhältnis zwischen oberem und unterem Meßpunkt, h = wahre Strahlenweglänge durch die Schicht)
die spektralen Extinktionskoeffizienten $a(\lambda)$ bestimmen. Dabei ist zu berücksichtigen, daß die optischen Dichten D des Spektrums auf dem Film der Beziehung $D = \ln I \cdot t$ genügen.

Aus einer Serie von 8 Messungen an 5 verschiedenen Tagen der Monate Mai/Juni 1975 ergaben sich für einige ausgewählte Wellenlängen die in der Tabelle 2 enthaltenen charakteristischen Extinktionskoeffizienten für die atmosphärische Bodenschicht über dem Stadtgebiet Leipzig im Sommer. Da der Trübungsfaktor als Quotient des realen Extinktionskoeffizienten zu dem einer Rayleigh-Atmosphäre definiert ist, erhält man aus den verfügbaren Werten die ebenfalls in Tabelle 2 enthaltenen spektralen Trübungsfaktoren, die naturgemäß mit zunehmender Wellenlänge anwachsen.

Tabelle 2

Wellenlängenspezifische Extinktionskoeffizienten $a(\lambda)$ und spektrale Trübungsfaktoren $\tau(\lambda)$ der atmosphärischen Bodenschicht über Leipzig, Mai/Juni 1975.
$a(\lambda)_R$: Extinkt.-Koeff. einer Rayleigh-Atmosphäre

λ :	331	339	492	553	634	[nm]
$a(\lambda)$:	5,92	6,08	6,46	6,54	5,21	$[10^{-4} m^{-1}]$
$a(\lambda)_R$:	1,010	0,492	0,195	0,121	0,070	$[10^{-4} m^{-1}]$
$\tau(\lambda)$:	5,9	12,4	33,2	54,0	74,4	

Legt man das erhaltene Maximum bei 550 nm in der ermittelten Extinktionsverteilung mit a (550 nm) = $6{,}5 \cdot 10^{-4}\ m^{-1}$ einer Modellberechnung der verursachenden Partikelverteilung zugrunde, so erhält man die in Abbildung 1 dargestellten Partikelverteilungen. Dabei wurde so vorgegangen, daß nach den Modellen von *Spänkuch* (1974) Medianwerte normierter logarithmisch-normalverteilter Partikelkollektive für verschiedene Streuparameter σ abgeschätzt wurden. Hieraus wurden dann Verteilungen bestimmt, die dem gemessenen Extinktionswert bei 550 nm entsprechen (*v. Hoyningen-Huene* 1977).

Abschließend sei noch erwähnt, daß die Methode sehr gut zur Bestimmung der kürzesten, den Meßpunkt erreichenden Wellenlänge des UV geeignet ist. Es zeigt sich, daß Wellenlängen < 300 nm in Leipzig im allgemeinen nur bei Sonnenhöhen > 40° bis zur Erdoberfläche durchdringen, während dies in der sauberen Atmosphäre des Küstengebietes (Zingst, Ostsee) schon bei Sonnenhöhen ≧ 20° der Fall ist. Mittlere Beziehungen zwischen Sonnenhöhe und der kürzesten nachweisbaren Wellenlänge für Leipzig und für Zingst sind aus der Abbildung 2 ersichtlich (*Baumgartl*, 1977).

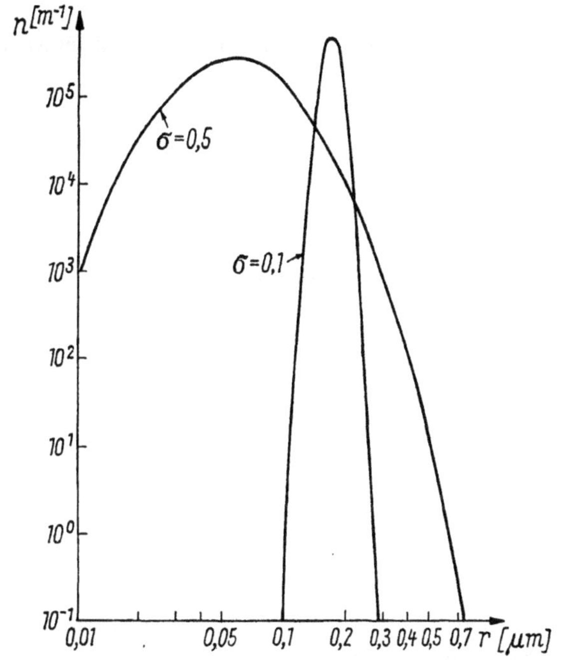

Abb. 1. Berechnete Partikelverteilungen für das Extinktionsmaximum bei 550 nm unter Vorgabe logarithmisch-normalverteilter Partikelkollektive mit verschiedenen Streuparametern.

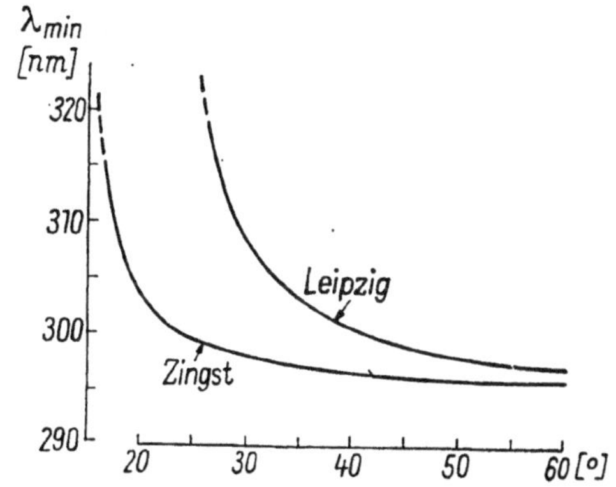

Abb. 2. Mittlere Beziehungen (Sommer) zwischen der kürzesten Wellenlänge der Sonnenstrahlung an der Erdoberfläche u. der Sonnenhöhe in einer sauberen (Zingst) u. einer verunreinigten Atmosphäre (Leipzig).

Literatur:

Andrejtschin, R. und *T. Kechlibarov:* Über die Messung der kurzwelligen Grenze der UV-Sonnenstrahlung. Comptes Rendus Bulgar. Acad. Sci. 16 (1963), 601–604

Baumgartl, M.: Untersuchungen der Beeinflussung der Sonnenstrahlung im uv- und angrenzenden sichtbaren Bereich durch eine Stadtatmosphäre. Geophysik u. Geologie (im Druck)

Foitzik, L. und *H. Hinzpeter:* Sonnenstrahlung und Lufttrübung. Akad. Verl.-Ges. Geest & Portig, Leipzig (1958)

v. Hoyningen-Huene, W.: Modellrechnungen zur optischen Wirkung von Partikelkollektiven (unveröff. Manuskript, 1977)

Lauscher, F.: Über den Trübungsfaktor. Meteor. Z. 47 (1930), 1–10

Linke, F.: Messung der Sonnenstrahlung bei vier Freiballonfahrten. Beitr. Phys. fr. Atm., Hergesell-Festschr. (1929), 176–184

Spänkuch, D.: Optische Eigenschaften logarithmisch-normalverteilter Partikelkollektive. Abh. Meteor. D. DDR (Nr. 109 (1974)

WMO: Urban Climates. WMO, Techn. Note Nr. 108 (1970)

Anschrift: Doz. Dr. sc. nat. Chr. Hänsel, Sektion Physik, Fachbereich Geophysik, der Karl-Marx-Universität, DDR-701 Leipzig, Talstraße 35

Druckfehlerberichtigung:

In Abbildung 1 muß die Größenangabe der Ordinate anstelle von n [m^{-1}] richtig

n [cm^{-3}]

heißen.

551.509.58
551.586:631.43

Тепло- влагоперенос в промерзающей почве и агрометеорологический прогноз

В. Н. Лыкосов Э. Г. Палагин

2 Рис.

Резюме: В работе проводится математическое моделирование процессов переноса в промерзающей почве. Формулируются соответствующие уравнения, и предлагается алгоритм их численного решения. Решение проводиться с учетом переменной во времени толщины снежного покрова. Даются примеры расчетов.

В прикладном отношении результаты ориентированы на область агрометеорологии.

Zusammenfassung: In der Arbeit erfolgt eine mathematische Modellierung der Prozesse dieses Transports in einem gefrierenden Boden. Es werden die entsprechenden Gleichungen formuliert und der Algorithmus ihrer Lösung vorgelegt. Die Lösung erfolgt unter Berücksichtigung der zeitlich veränderlichen Dicke der Schneedecke. Es werden Beispiele von Berechnungen angegeben.

In Bezug auf die Anwendung sind die Ergebnisse auf das Gebiet der Agrarmeteorologie orientiert.

Термический и влажностный режим почвы определяет условия перезимовки озимых культур и, следовательно, степень их повреждения к моменту весенней вегетации. Так например, опускание температуры на уровне узла кущения ниже некоторого предела ведет к его повреждению или гибели. Накопление влаги в верхних слоях почвы может повести к ее вспучиванию и выпиранию узлов кущения на порехность, что также приводит к возникновению опасных для растений ситуаций. Длительное пребывание под мошным снежным покровом при температурах близких к нулю °С будет способствовать выпреванию и так далее. Для того, чтобы иметь представление о состоянии озимых посевов к концу зимнего сезона необходимо иметь регулярно поступающую информацию о температуре и влажности в почве. Однако, измерение этих характеристик проводиться далеко не всегда и не везде. Поэтому возникает вопрос о создании методики, позволяющей рассчитывать поля температуры и влажности как в диагностическом, так и прогностическом плане, Решению этого вопроса и посвящена предлагаемая работа.

Процессы обмена, протекающие в почве, главным образом обусловлены воздействием всей совокупности метеорологических факторов. Поэтому в принципе, формулируя задачу, следует рассматривать систему атмосфера — снег — почва. Однако в зимний период, поскольку преобладающей является нейтральная стратификация, допустимо считать температуру поверхности в среднем равной температуре воздуха на уровне метеобудки $T_B(t)$.

В дальнейшем будем рассматривать одномерный случай, ибо вертикальные градиенты температуры и влажности, как правило, значительно больше горизонтальных. Зависимость от горизонтальных координат в этом случае может быть учтена параметрически. Отметим еще, что в снеге массообмена практически не происходит, а термические процессы могут быть описаны с помощью уравнения теплопроводности.

$$c_s(z,t)\,\frac{\partial T_s}{\partial t}=\frac{\partial}{\partial z}\,\lambda_s(z,t)\,\frac{\partial T_s}{\partial z}\,. \tag{1}$$

(Здесь начало координат расположено на поверхности почвы и ось z направлена вниз). На поверхности снега имеем:

$$T_s\big|_{z=-l(t)}=T_B(t)\,. \tag{2}$$

Прежде чем продолжить математическую формулировку задачи, необходимо вывести уравнения переноса тепла и влаги в промерзающей почве.

Последнюю при этом будет трактовать как континуум, состоящий из скелета минерально-органического происхождения, а также льда, жидкой воды и паровоздушной смеси, заполняющих поры. Таким образом ее плотность $\varrho=\sum_{i=1}^{5}\varrho_i$ (индексы $i=1, 2, 3, 4, 5$ соответственно отвечают льду, скелету, воде, воздуху и пару). Скорость центра масс этой систены $v=\sum_{i=1}^{5}\alpha_i v_i/\varrho$. Если пренебречь незначительными деформациями почвы, то связанная с ней система координат будет неподвижна в пространстве и поэтому $v_1=v_2=0$. Тогда без существенного ущерба для точности можно полагать, что массоперенос осуществляется только за счет движения жидкой фазы, т.е. $v=\alpha_3 v_3/\varrho$.

Уравнения баланса массы для льда и воды соответственно запишуться:

$$\frac{\partial\alpha_1}{\partial t}=I_1, \tag{3}$$

$$\frac{\partial\alpha_3}{\partial t}=-\frac{\partial\alpha_3 v_2}{\partial z}+I_3\,. \tag{4}$$

Ясно, что $I_1 = -I_3$, ибо жидкость переходит в лед. В ведя лиффузионный поток $J_3 = \alpha_3 (v_3 - v)$, будем иметь: $\alpha_3 v_3 = J_3 + \alpha_3 v$, где $\alpha_3 v$ представляет совой поток, обусловленный вынужденной конвекцией. В пористых средах при отсутствии фильтрационных движений конвективной составляющей можно пренебречь [1]. Тогда, используя весовые влажности для воды $w_в = \alpha_3/\gamma$ и льда $w_Л = \alpha_1/\gamma$ можно [3] переписать в виде:

$$(5) \qquad \frac{\partial w_B}{\partial t} = -\frac{1}{\gamma}\frac{\partial J_3}{\partial z} - \frac{\partial w_Л}{\partial t}$$

Уравнение движения и баланса внутренней энергии при протекании чисто диффузионных процессов соответственно имеют вид [2]:

$$(6) \qquad \frac{\partial p}{\partial z} = g\varrho$$

(механическое равновесие)

$$(7)^{1)} \qquad \frac{\partial \varrho u}{\partial t} = -\frac{\partial J_u}{\partial z}.$$

Отметим, что J_u служит точным определением потока тепла т.е. $J_q = J_u$ [3]. Введя удельную энтальпию $h = u + p/\varrho$ и учитывая, что $\frac{\partial p}{\partial t} = 0$, из (7) получим:

$$(8) \qquad \frac{\partial \varrho h}{\partial t} = -\frac{\partial J_u}{\partial z}.$$

Но $\varrho h = \sum_{i=1}^{5} \alpha_i h_i$. Тогда после несложных преобразований, используя (3), (5), уравнение (8) можно представить в виде:

$$(9) \qquad \varrho c_p \frac{\partial T}{\partial t} = -\frac{\partial}{\partial z}\left(J_q - \sum_{i=1}^{5} h_i J_i\right) + L\gamma \frac{\partial w_Л}{\partial t} - \sum_{i=1}^{5} c_{pi}(J_i \vec{\nabla})T.$$

Здесь $L = h_3 - h_1$; $\varrho c_p = \sum_{i=1}^{5} \alpha_i \frac{\partial h_i}{\partial T}$. Сравним два последных члена в уравнении (9), записав их отношение в виде характерных значений входящих в них величин (отмечены индексом «О»). Тогда $\frac{\Delta T_0 J_{0i} c_{pi}}{z_0 \cdot L} \Big/ \left(\frac{\partial w_Л}{\partial t}\right)_0$. Если даже весь поток жидкости, входящий в единичный об'ем, превратится в лед, чему соответствует $\frac{J_{0i}}{z_0} \Big/ \left(\frac{\partial w_Л}{\partial t}\right)_0 \sim$ ~ 1 то и тогда $\frac{\Delta T_0 c_{pi}}{L} \ll 1$. Отсюда следует, что последним слагаемым в (9) можно пренебречь.

Для определения явного вида потоков J_3, J_9, входящих в (5) и (9), воспользуемся методами термодинамики необратимых процессов. Источник энтропии применительно к данной задаче имеет вид:

$$(10) \qquad \sigma = J_q \cdot X_q + \sum_{i=1}^{5} J_i X_i.$$

Термодинамическая сила, сопряженная потоку тепла $X_q = \frac{\partial}{\partial z}\left(\frac{1}{T}\right)$, а потоку массы $X_i = \frac{g}{T} - \frac{\partial}{\partial z}\left(\frac{\mu_i}{T}\right)$, где $\mu_i = \mu_i(T, p, \alpha_i)$ — химический потенциал (парциальная удельная функция Гиббса) почвы. Как известно: $\frac{\partial \mu_i}{\partial t} = h_i \frac{\partial}{\partial z}\left(\frac{1}{T}\right)$; $\frac{\partial \mu_i}{\partial p} = \frac{1}{\varrho_i}$. Тогда, учитывая (6), получим:

$$(11) \qquad X_i = -h_i \frac{\partial}{\partial z}\left(\frac{1}{T}\right) + \frac{1}{T}\left[\left(1 - \frac{\varrho}{\varrho_i}\right) g - \left(\frac{\partial \mu_i}{\partial z}\right)_{P,T}\right].$$

Тогда источник энтропии может быть преобразован:

$$(10') \qquad \sigma = J_q' X_q + \sum_{i=1}^{5} J_i X_i'.$$

В (10′) введен новый поток тепла $J_q' = J_q - \sum_{i=1}^{5} h_i J_i$ которому как и ранее, соответствует сила X_q, а потоку массы теперь сопряжена сила X_i'. Поскольку $\sum_{i=1}^{5} J_i = 0$, то, исключая $J = -\sum_{i=2}^{5} J_i$, будем иметь:

$$(10'') \qquad \sigma = (J_q' \cdot X_q) + \sum_{i=2}^{5} J_i \cdot X_i''.$$

где

$$X_i'' = \frac{1}{T}\left[\left(\frac{1}{\varrho_1} - \frac{1}{p_i}\right) g\varrho - \frac{\partial}{\partial z}(\mu_i - \mu_1)_{P,T}\right].$$

---

1) Здесь также использованы сделанные ранее допущения.

Связь между потоками и силами дается выражениями:

$$J'_q = L_{qq} \cdot X_q + \sum_{i=2}^{5} L_{qi} \cdot X''_i ; \quad J_i = L_{iq} \cdot X_q + \sum_{k=2}^{5} L_{ik} \cdot X''_k ;$$

L_{qq}, L_{qi}, L_{iq}, L_{ik} -феномелогические коэффициенты для которых справедливы соотношения Онсагера $L_{iq} = L_{qi}$; $L_{ik} = L_{ki}$. Теперь можно записать:

$$J'_q = -\frac{L_{qi}}{T^2}\frac{\partial T}{\partial z} + \frac{1}{T}\sum_{i=2}^{5} L_{qi}\left[\left(\frac{1}{\varrho_1} - \frac{1}{\varrho_i}\right)\varrho g - \frac{\partial}{\partial z}(\mu_i - \mu_i)_{P,T}\right] \tag{12}$$

$$J_3 = -\frac{L_{3q}}{T^2}\frac{\partial T}{\partial z} + \frac{1}{T}\sum_{k=2}^{5} L_{3k}\left[\left(\frac{1}{\varrho_1} - \frac{1}{\varrho_k}\right)\varrho g - \frac{\partial}{\partial z}(\mu_k - \mu_1)_{P,T}\right].$$

В (12) разность потенциалов есть функция α_i $(i = 2, \ldots, 5)$.
Поэтому:

$$\sum_{i=2}^{5}\frac{L_{qi}}{T}\sum_{k=2}^{5}\frac{\partial(\mu_i - \mu_1)_{P,T}}{\partial\alpha_k}\frac{\partial\alpha_k}{\partial z} = \sum_{k=2}^{5}\underbrace{\left(\sum_{i=2}^{5}\frac{L_{qi}}{T}\frac{\partial(\mu_i - \mu_1)_{P,T}}{\partial\alpha_k}\right)}_{D_{q,k}}\frac{\partial\alpha_k}{\partial z} = \sum_{k=2}^{5} D_{qk}\frac{\partial\alpha_k}{\partial z}.$$

Аналогичным образом:

$$\sum_{k=2}^{5}\frac{L_{3k}}{T}\sum_{j=2}^{5}\frac{\partial(\mu_k - \mu_1)_{P,T}}{\partial\alpha_j}\frac{\partial\alpha_j}{\partial z} = \sum_{j=2}^{5}\underbrace{\left(\sum_{k=2}^{5}\frac{L_{3k}}{T}\frac{\partial(\mu_k - \mu_1)_{P,T}}{\partial\alpha_j}\right)}_{D_{3,j}}\frac{\partial\alpha_j}{\partial z} = \sum_{j=2}^{5} D_{2j}\frac{\partial\alpha_j}{\partial z}.$$

Если дополнительно ввести обозначения:

$$\lambda = \frac{L_{qq}}{T^2}, \quad \lambda_q = \frac{L_{3,p}}{T^2}; \quad k_g = \sum_{k=2}^{5}\frac{L_{qi}}{T}\left(\frac{1}{\varrho_1} - \frac{1}{\varrho_i}\right)\varrho_i; \quad k_3 = \sum_{k=2}^{5}\frac{L_{3k}}{T}\left(\frac{1}{\varrho_i} - \frac{1}{\varrho_i}\right)\varrho$$

то (12) перепишутся в виде

$$\left.\begin{aligned} J'_q &= -\lambda\frac{\partial T}{\partial z} - \sum_{k=2}^{5} D_{qk}\frac{\partial\alpha_k}{\partial z} + k_g g \\ J_3 &= -\lambda_q\frac{\partial T}{\partial z} - \sum_{j=2}^{5} D_{3j}\frac{\partial\alpha_j}{\partial z} + k_3 g \end{aligned}\right\}. \tag{12'}$$

Поскольку влияние перекрестного эффекта переноса массы на J_3 мало, то в (12) следует оставить слагаемые с $k = j = 3$. Тогда (5) и (9) можно записать следующим образом:

$$\frac{\partial w_B}{\partial t} = \frac{\partial}{\partial z}\left(D_{33}\frac{\partial w_B}{\partial z} + \frac{\lambda_g}{\gamma}\frac{\partial T}{\partial z}\right) - \frac{g}{\gamma}\frac{\partial k_3}{\partial z} - \frac{\partial w_A}{\partial t} \tag{13}$$

$$\varrho c_p\frac{\partial T}{\partial t} = \frac{\partial}{\partial z}\left(\lambda\frac{\partial T}{\partial z} + D_{q3}\gamma\frac{\partial w_B}{\partial z}\right) - g\frac{\partial k_g}{\partial z} + L\gamma\frac{\partial w_A}{\partial t}. \tag{14}$$

В (13) и (14) в весьма общей форме учитываются основные факторы, обуславливающие транспорт тепла и влаги при промерзании почвы. Однако не все они равноценны по значению. Так например, энергия поверхностного взаимодействия влаги с почвой меняется в диапазоне $0 \div 300$ дж/г, что соответствует давлению $0 \div 3000$ атм [4]. В сравнении с этими величинами вклад сил тяжести в ненасщенных почвах мал. Но при фильтрации в почву, проходящей под слоем воды, учет гравитационных эффектов весьма существенен. В данном случае мы рассматриваем миграцию в зимний период, и будем исключать случаи интенсивного снеготаяния. Поэтому члены, отражающие влияние силы тяжести в (13)–(14) можно опустить. Тогда выведенные нами уравнения по внешнему виду формально совпадают с уравнениями, приведенными в [1]. Но смысл и внутренняя структура коэффициентов отличны от тех, которые использовались авторами указанной монографии. Это имеет принципиальное значение при разработке соответствующей методики их экспериментального определения, но с точки зрения расчетной процедуры не играет роль, что позволяет использовать накопленный к настоящему моменту эмпирический материал. Именно это обстоятельство побудило сохранить внешний вид уравнений.

Отметим, что для зоны плюсовых температур все положения, лежащие в основе вывода (13)–(14) остаются в силе. В самих уравненииних изменяются лишь численные значения коэффициентов, и источники следует положить равными нулю. Тогда окончательно исходные уравнения запишутся в виде:

$$\frac{\partial w_j}{\partial t} = \frac{\partial}{\partial z}\left[D_j\left(\frac{\partial w_j}{\partial z} + \delta^B_{Tj}\frac{\partial T_j}{\partial z}\right)\right] - \frac{\partial w_A}{\partial t}\delta_{1j} \tag{15}$$

$$c_j\frac{\partial T_j}{\partial t} = \frac{\partial}{\partial z}\left[\lambda_j\left(\frac{\partial T_j}{\partial z} + \delta^T_{Bj}\frac{\partial w_j}{\partial z}\right)\right] + L\gamma\frac{\partial w_A}{\partial t}\delta_{1j} \tag{16}$$

Здесь обозначено: $D_j \equiv D_{33}$; $\delta^B_{Tj} = \lambda_q/\gamma D$; $\delta^T_{Bj} = \gamma D_{q3}/\lambda_j$;

К системе уравнений (1), (15), (16) необходимо добавить граничные и начальные условия. Очевидно, что на поверхности $z = 0$ отсутствует перетекание влаги в снег, и имеет место равенство температур и потоков тепла:

$$T|_{z=-0} = T_1|_{z=+0}; \; \lambda_s(z,t)\frac{\partial T_s}{\partial z}\bigg|_{z=-0} = \lambda_1(z_1 t)\left(\frac{\partial T_1}{\partial z} + \delta^B_{T_1}\frac{\partial w_1}{\partial z}\right|_{z=+0} D, \gamma\left(\frac{\partial w_1}{\partial z} + \delta^T_{B_1}\frac{\partial T_1}{\partial z}\right)\bigg|_{z=+0} = 0. \tag{17}$$

На границе раздела мерзлая-талая почва при $z = h$ имеет место баланс тепла и влаги. Первое из них есть следствие равенства потока тепла, отводимого в мерзлую зону J_{q1}, сумме, которая состоит из потока тепла со стороны талой почвы J_{q2}, и тепла, выделяющегося за единицу времени в единице об'ема при движении нулевой изотермы $L\gamma\left[w_2 - w_1(0)\,\frac{dh}{dt}\right]$, где $w_1(0)$ – количество жидкой влаги, незамерзающей при 0 °C. Тогда

$$J_{q1}\,|_{z=h(t)-0} = -J_{q2}\,|_{z=h(t)+0} + L\gamma\,[w_2 - w_1(0)\,\frac{dh}{dt}\,. \tag{18}$$

Второе можно сформулировать, исходя из следующих соображений. Общая скорость поступления жидкости к границе равна $\left|\frac{dh}{dt} - v_3\right|_{z=h(t)+0}$, а поток

$$\left|\gamma w_2\left(\frac{dh}{dt} - v_3\right)\right|_{z=h(t)+0} = \left|-J_2 + \gamma w_2\,\frac{dh}{dt}\right|_{z=h(t)+0}.$$

Отводиться сумма: жидкая влага в количестве $\left|-J_1 + \gamma w_1(0)\,\frac{dh}{dt}\right|_{z=h(t)-0}$ и лед $\gamma w_\Lambda\,\frac{dh}{dt}$. Тогда с учетом того, что общее количество влаги на границе не терпит разрыва, получим:

$$J_1|_{z=h(t)-0} J = {}_2|_{z=h(t)+0}\,.$$

Кроме того, на той же границе выполняются равенства:

$$T_1|_{z=h(t)-0} = T_2|_{z=h(t)+0}\,, \quad w|_{z=h(t)-0} = W_2|_{z=h(t)+0} \tag{20}$$

$$w = w_1 + w_\Lambda$$

Нижней границей $z = H$ будет считаться уровень затухания сезонных колебаний температуры. На ней выполняются условия:

$$T|_{z=H} = T_H\,, \quad w_2|_{z=H} = w_H\,. \tag{21}$$

Если влажность меньше влажности разрыва капилляров, то

$$\left|\lambda_q\,\frac{\partial T}{\partial z} + D_{33}\,\frac{\partial w_B}{\partial z}\right|_{z=H} = 0\,. \tag{21'}$$

Наконец, может произойти замена второго из условий (21) на (21′), если влажность в связи с опусканием уровня грунтовых вод сильно меняется.

Начальные условия (по всей области):

$$T|_{t=0} = T^0(z)\ ; \quad w_j|_{t=0} = w_j^0(z) \tag{22}$$

В результате решения (1), (2), (15)–(22) могут быть найдены профили температуры и общей влажности для любого момента времени. Причем последняя величина в мерзлой зоне определяется как сумма предшествующей полной влажности и полученного на расчетном шаге притока жидкой влаги. Для выделения твердой фазы и определения вида $w_1(0)$ используем эмпирическую зависимость содержания незамерзщей воды от температуры $w_1 = w_{\max}(T)$[1] (при $T \leqq 0$ °C).

Вводим: $\xi = I + z/l(t)$; $0 \leqq \xi \leqq 1$, если $-l(t) \leqq z < 0$

$\xi = I + z/h(t)$; $1 \leqq \xi \leqq 2$, если $0 \leqq z < h(t)$

$\xi = 2 + (z - h(t))/(H - h(t))$; $2 \leqq \xi \leqq 3$, если $h(t) \leqq z \leqq H$.

Тогда (1), (2), (15)–(22) перепишутся в виде:

$$c_s\,\frac{\partial T_s}{\partial t} = \frac{\partial}{\partial \xi}\,K_1^s\,\frac{\partial T_s}{\partial \xi} + K_2^s c_s\,\frac{\partial T_s}{\partial \xi} \tag{23}$$

$$c\,\frac{\partial T}{\partial t} = \frac{\partial}{\partial \xi}\left(K_{11}\,\frac{\partial T}{\partial \xi} + K_{12}\,\frac{\partial w}{\partial \xi}\right) + cK_2\,\frac{\partial T}{\partial \xi} + L\gamma\left(\frac{\partial w_\Lambda}{\partial t} + K_2\,\frac{\partial w_\Lambda}{\partial \xi}\right) \tag{24}$$

$$\frac{\partial w}{\partial t} = \frac{\partial}{\partial \xi}\left(K_{21}\,\frac{\partial T}{\partial \xi} + K_{22}\,\frac{\partial w}{\partial \xi}\right) + K_2\,\frac{\partial w}{\partial \xi} - \left(\frac{\partial w_\Lambda}{\partial t} - K_2\,\frac{\partial w_\Lambda}{\partial \xi}\right) \tag{25}$$

$$T_s\,|_{\xi=0} = T_B\,(t) \equiv T_s^*(t) \tag{26}$$

$$T_s|_{\xi=-0} = T|_{\xi=+0}\ ; \quad \frac{\lambda_s}{l}\,\frac{\partial T_s}{\partial \xi}\bigg|_{\xi=1-0} = \frac{\lambda}{h}\left(\frac{\partial T}{\partial \xi} + \delta_T^B\,\frac{\partial w}{\partial \xi}\right)\bigg|_{\xi=1+0}; \quad \frac{D_1}{h}\,\gamma\left(\frac{\partial w}{\partial \xi} + \delta_B^T\,\frac{\partial T}{\partial \xi}\right)\bigg|_{\xi=1+0} = 0 \tag{27}$$

$$\left.\begin{aligned} &T|_{\xi=2-0} = T|_{\xi=2+0}\ ; \quad w|_{\xi=2-0} = w_2|_{\xi=2+0} \\ &\frac{\lambda}{h}\left(\frac{\partial T}{\partial \xi} + \delta_T^B\,\frac{\partial w}{\partial \xi}\right)\bigg|_{\xi=2-0} - \frac{\lambda}{H-h}\left(\frac{\partial T}{\partial \xi} + \delta_T^B\,\frac{\partial w}{\partial \xi}\right)_{\xi=2+0} = L\gamma w_\Lambda|_{\xi=2}\,\frac{dh}{dt} \end{aligned}\right\} \tag{28}$$

$$T|_{\xi=3} = T_H\ ; \quad w|_{\xi=3} = w_H \tag{29}$$

---

[1]) Каждая кривая отвечает какому-либо конкретному виду почв.

(Для талой почвы алгоритм сохраняется, упрощаясь за счет того, что $w_A=0$. Поэтому индекс «j» здесь и далее опускается). В (23)–(25) введены обозначения: $K_1^s=\lambda_s/l^2$; $K_{11}=\lambda/\tilde{h}^2$;

$$K_2^s=\frac{dl}{dt}(\xi-1)/l\ ;\quad K_{12}=\delta_T^B\cdot\lambda/\tilde{h}^2\ ;\quad K_{21}=\delta_B^T\cdot D/\tilde{h}^2\ ;\quad K_{22}=D/\tilde{h}^2$$

$$\tilde{h}=\begin{cases}h, & \text{если } 1<\xi<2,\\ H-h, & \text{если } 2<\xi<3,\end{cases}\qquad K_2=\frac{dh}{dt}\cdot\begin{cases}(\xi-1)/h & \text{, если } 1<\xi<2\\ (3-\xi)/(H-h), & \text{если } 2<\xi<3\,.\end{cases}$$

Для решения задачи (23)–(29) используем конечно-разностный метод, Вводим в рассмотрение сетку по переменной t: ($t_j=t\cdot\Delta t$; $j=0,1,2,\ldots$; Δt – шаг по времени). На каждом шаге по времени $t_j<t<t_{j+1}$ систему уравнений (23)–(25) расщепим на две подсистемы. На первом полушаге $t_j=t=t_{j+\frac{1}{2}}$ будем решать задачу диффузии тепла и влаги з почве и снеге.

$$(30)\qquad (c_s)^j\,\frac{T_s^{j+\frac{1}{2}}-T_s^j}{\Delta t}=\frac{\partial}{\partial\xi}(K_1^s)^j\,\frac{\partial T^{j+\frac{1}{2}}}{\partial\xi}+(K_2^s)^j\,(c_s)^j\left(\frac{\partial T_s}{\partial\xi}\right)^{j+\frac{1}{2}}$$

$$(31)\qquad c^j\,\frac{T^{j+\frac{1}{2}}-T^j}{\Delta t}=\frac{\partial}{\partial\xi}\left(K_{11}^j\,\frac{\partial T^{j+\frac{1}{2}}}{\partial\xi}+K_{12}^j\,\frac{\partial w^{j+\frac{1}{2}}}{\partial\xi}\right)+(cK_2)^j\,\frac{\partial T^{j+\frac{1}{2}}}{\partial\xi}-L\gamma K_2^j\left(\frac{\partial w_A}{\partial\xi}\right)^{j+\frac{1}{2}}$$

$$(32)\qquad \frac{w^{j+\frac{1}{2}}-w^j}{\Delta t}=\frac{\partial}{\partial\xi}\left(K_{21}^j\,\frac{\partial T^{j+\frac{1}{2}}}{\partial\xi}+K_{22}^j\,\frac{\partial w^{j+\frac{1}{2}}}{\partial\xi}\right)+K_2^j\,\frac{\partial w_A^{j+\frac{1}{2}}}{\partial\xi}$$

$$(33)\qquad w_A^{j+\frac{1}{2}}=w_A^j$$

$\varphi^j=\varphi(t^j)/\varphi(\varphi$ — любая из рассматриваемых функций). Начальные условия при $t=t_j$:

$$(34)\qquad T_s=T_s^j;\ T=T^j;\ w=w^j;\ w_A=w_A^j\,.$$

Граничные условия (26), (27), (29) записываются для момента $t=t_{j+\frac{1}{2}}$.

На втором полушаге $t_{j+\frac{1}{2}}\leqq t\leqq t_{j+1}$ решается задача адаптации профилей, несколько рассогласованных на предыдущем полушаге. Сначала рассмотрим первый полушаг, т.е. этап расчета диффузии. Вводим векторы $\Phi=\begin{pmatrix}T\\ w\end{pmatrix}$, $R=\Delta t K_2^j\left(\frac{\partial W_A}{\partial\xi}\right)^{j+\frac{1}{2}}\begin{pmatrix}-L_\gamma\\ 1\end{pmatrix}$ и матрицы $K_1=\Delta t\begin{pmatrix}K_{11}^j & K_{12}^j\\ K_{21}^j & K_{22}^j\end{pmatrix}$, $S=\begin{pmatrix}c^j & 0\\ 0 & 1\end{pmatrix}$. Тогда уравнения (31), (32) могут быть переписаны в матричном виде:

$$(35)\qquad S\left(\Phi^{j+\frac{1}{2}}-\Phi^j\right)=\frac{\partial}{\partial\xi}K_1\,\frac{\partial\Phi^{j+\frac{1}{2}}}{\partial\xi}+K_2^jS\,\frac{\partial\Phi^{j+\frac{1}{2}}}{\partial\xi}+R\,.$$

Разностную аппроксимацию по ξ для (35) построим с помощью интегро-интерполяционного метода [5]. Введем в рассмотрение систему точек по ξ (ξ_m), располагая их в порядке возрастания индекса «m»: $\xi_0=0,\ldots,\xi_{m_1}=1,\ldots,\xi_{m_2}=2,\ldots,\xi_{m_3}=3$. Для каждого интервала (ξ_m, ξ_{m+1}) рассмотрим среднюю точку $\xi_{m+\frac{1}{2}}=\frac{1}{2}(\xi_m+\xi_{m+1})$.

Пусть $m\neq m_2$. Проинтегрируем (35) по ξ от $\xi_{m+\frac{1}{2}}$ до $\xi_{m+\frac{1}{2}}$ и заменим затем интегралы квадратурными формулами.

Будем иметь:

$$(36)\qquad \Delta\xi_mS_m\left(\Phi_m^{j+\frac{1}{2}}-\Phi_m^j\right)=I_{m+\frac{1}{2}}-I_{m+\frac{1}{2}}+(K_2^j)_mS_m\left(\Phi_{m+\frac{1}{2}}^{j+\frac{1}{2}}-\Phi_{m+\frac{1}{2}}^{j+\frac{1}{2}}\right)+R_m\Delta\xi_m$$

где

$$(37)\qquad \frac{\partial\Phi^{j+\frac{1}{2}}}{\partial\xi}=K_1^{-1}I\ ;\quad \Phi_{m\pm\frac{1}{2}}^{j+\frac{1}{2}}=\frac{1}{2}\left(\Phi_m^{j+\frac{1}{2}}+\Phi_{m+\frac{1}{2}}^{j+\frac{1}{2}}\right)$$

$$\Delta\xi_m=\xi_{m+\frac{1}{2}}-\xi_{m-\frac{1}{2}}\,,\quad \varphi_m=\varphi(\xi_m)\,.$$

Интегрируя первое из соотношений от ξ_m до ξ_{m+1}, получим

$$(38)\qquad \Phi_{m+1}^{j+\frac{1}{2}}-\Phi_m^{j+\frac{1}{2}}\approx\int_{\xi_m}^{\xi_{m+1}}K_1^{-1}d\xi\cdot I_{m+\frac{1}{2}}\,.$$

Подставляя (38) в (36) получим следующие разностные уравнения:

$$\Delta\xi_m S_m\left(\Phi_m^{j+\frac{1}{2}}-\Phi_m^j\right)=\frac{1}{\Delta\xi_{m+\frac{1}{2}}}(K_1)_{m+\frac{1}{2}}\left(\Phi_{m+1}^{j+\frac{1}{2}}-\Phi_m^{j+\frac{1}{2}}\right)-\frac{1}{\Delta\xi_{m-\frac{1}{2}}}(K_1)_{m+\frac{1}{2}}\left(\Phi_m^{j+\frac{1}{2}}-\Phi_{m-1}^{j+\frac{1}{2}}\right)+\frac{1}{2}(K_2)_m S_m\left(\Phi_{m+1}^{j+\frac{1}{2}}-\Phi_{m-1}^{j+\frac{1}{2}}\right)+R_m\Delta\xi_m \tag{39}$$

$$(K_1)_{m+\frac{1}{2}}=\Delta\xi_{m+\frac{1}{2}}\left(\int\limits_{\xi_m}^{\xi_{m+1}}K_1^{-1}d\xi\right)^{-1}$$

$$\int\limits_{\xi_m}^{\xi_{m+1}}K_1^{-1}d\xi=\frac{1}{2}\left[(K_1^{-1})_m+(K_1^{-1})_{m+1}\right](\xi_{m+1}-\xi_m)\,.$$

При $m=m_2$ система (39) модифицируется таким образом, чтобы удовлетворялись условия (28). Аналогичным образом проводилась конечно-разностная аппроксимация по ξ и для (30). В итоге получена следующая система:

$$(C_s)^j\left[\left(T_s^{j+\frac{1}{2}}\right)_m-(T_s^j)_m\right]\Delta\xi_m=(K_1^s)^j_{m+\frac{1}{2}}\left[\left(T_s^{j+\frac{1}{2}}\right)_{m+1}-\left(T_s^{j+\frac{1}{2}}\right)_m\right]\Big/\Delta\xi_{m+\frac{1}{2}}-(K_1^s)^j_{m-\frac{1}{2}}\left[\left(T_s^{j+\frac{1}{2}}\right)_m-\left(T_s^{j+\frac{1}{2}}\right)_{m-1}\right]\Big/\Delta\xi_{m-\frac{1}{2}}+\frac{1}{2}(K_2^s)^j_m\left[\left(T_s^{j+\frac{1}{2}}\right)_{m+1}-\left(T_s^{j+\frac{1}{2}}\right)_{m-1}\right](c_s)^j_m \tag{40}$$

$$\times(K_1^s)^j_{m+\frac{1}{2}}\cong\Delta\xi_{m+\frac{1}{2}}\left(\int\limits_{\xi_m}^{\xi_{m+1}}\frac{d\xi}{(K_1^s)^j}\right)^{-1};\quad(m\neq 0,m_1)\,.$$

Схема реализации численного алгоритма такова. Вначале методом скалярной прогонки [6] решается (40) с условиями:

$$(T_s)_0^{j+\frac{1}{2}}=T_s^*\left(t^{j+\frac{1}{2}}\right);\quad(T_s)_{m_1}^{j+\frac{1}{2}}=T_{m_1}^j\,. \tag{41}$$

Затем, используя метод матричной прогонки [5], ищем решение системы (39) с условиями (42), аппроксимиру ющими (27)

$$(\lambda_s^j)_{m_1-\frac{1}{2}}\left[\left(T_s^{j+\frac{1}{2}}\right)_{m_1}-\left(T_s^{j+\frac{1}{2}}\right)_{m_1-1}\right]\Big/\left(l\cdot\Delta\xi_{m_1-\frac{1}{2}}\right)=(\lambda^j)_{m_1+\frac{1}{2}}\times\left(T_{m_1+1}^{j+\frac{1}{2}}-T_{m_1}^{j+\frac{1}{2}}\right)\Big/\left(h^j\cdot\Delta\xi_{m_1+\frac{1}{2}}\right)+(\lambda\delta_T^B)^j_{m_1+\frac{1}{2}}\left(w_{m_1+1}^{j+\frac{1}{2}}-w_{m_1}^{j+\frac{1}{2}}\right)\Big/\left(h^j\cdot\Delta\xi_{m_1+\frac{1}{2}}\right)(\gamma D)^j_{m_1+\frac{1}{2}}\left(w_{m_1+1}^{j+\frac{1}{2}}-w_{m_1}^{j+\frac{1}{2}}\right)\Big/\left(h^j\Delta\xi_{m_1+\frac{1}{2}}\right)+(D\delta_B^T)^j_{m_1+\frac{1}{2}}\left(T_{m_1+\frac{1}{2}}^{j+\frac{1}{2}}-T_{m_1}^{j+\frac{1}{2}}\right)\Big/\left(h^j\Delta\xi_{m_1+\frac{1}{2}}\right)=0 \tag{42}$$

и условиями:

$$T_{m_3}^{j+\frac{1}{2}}=T_H\left(t^{j+\frac{1}{2}}\right);\quad w_{m_3}^{j+\frac{1}{2}}=w_H\left(t^{j+\frac{1}{2}}\right). \tag{43}$$

На втором полушаге задача согласования решается лишь при $m_1\leqq m\leqq m_2$ и имеет вид:

$$(c)^j_m\,\frac{T_m^{j+1}-T_m^{j+\frac{1}{2}}}{\Delta t}=(L\gamma)_m\,\frac{(w_A)_m^{j+1}-(w_A)_m^{j+\frac{1}{2}}}{\Delta t} \tag{44}$$

$$\frac{w_m^{j+1}-w_m^{j+\frac{1}{2}}}{\Delta t}=-\frac{(w_A^{j+1})_m-\left(w_A^{j+\frac{1}{2}}\right)_m}{\Delta t} \tag{45}$$

С начальными условиями:

$$T_m=T_m^{j+\frac{1}{2}};\quad w_m=w_m^{j+\frac{1}{2}}\left(npu\ t=t_{j+\frac{1}{2}}\right);\quad(w_A)_m=\left(w_A^{j+\frac{1}{2}}\right)_m. \tag{46}$$

Заметим, что вторая задача решается только в том случае, если либо $w_m^{j+\frac{1}{2}}>w_{\max}\left(T_m^{j+\frac{1}{2}}\right)$, $T_m^{j+\frac{1}{2}}<0$, либо $T_m^{j+\frac{1}{2}}>0$ и $\left(w_A^{j+\frac{1}{2}}\right)_m>0$. При этом, если имеет место первое условие, то предполагается, что избыток влаги превращается в лед, т.е. система (44), (45) замыкается уравнением $w_m^{j+1}=w_{\max}(T_m^{j+1})$. В противном случае часть льда тает, так что температура изменяется до нуля. Тогда (44), (45) замыкается соотношением $T_m^{j+1}=0$.

Если ни одно из указанных условий не выполняется, то согласование не проводиться и значения зависимых величин переносятся с $\left(j+\frac{1}{2}\right)$ -го шага на $(j+1)$ — ый. Заметим, что в вышеприведенных формулах значение h предполагется известным с прошлого шага t . Лосле того как решение в $(j+1)$ — ом шаге будет получено, находиться значение L. с этой целью (28) аппроксимируется следующим разностным уравнением:

$$(47)\qquad (L\gamma)_{m_2}(w_\Lambda^{j+1})_{m_2}\frac{h^{j+1}-h^j}{\Delta t}=-\left\{\frac{(\lambda)^j_{m_2+\frac{1}{2}}T^{j+1}_{m_2+1}}{(H-h^j)\,\Delta\xi_{m_2+\frac{1}{2}}}+\frac{(\lambda\delta^B_T)^i_{m_2+\frac{1}{2}}(w^{j+1}_{m_2+1}-w^{j+1}_{m_2})}{(w-h^j)\,\Delta\xi_{m_2+\frac{1}{2}}}+\frac{\lambda^j_{m_2-\frac{1}{2}}T^{j+1}_{m_2-1}}{h^j\Delta\xi_{m_2-\frac{1}{2}}}-\frac{(\lambda\delta^T_B)^j_{m_2-\frac{1}{2}}(w^{j+1}_{m_2}-w^{j+1}_{m_2-1})}{h^j\,\Delta\xi_{m_2-\frac{1}{2}}}\right\}$$

При построении (47) учитывалось, что $T_{m_2}=0$ при всех j. При расчетах использовались следующие значения коэффициентов. Удельная теплоемкость снега равна 0,5 кал/г · град, а об'емная $c_s(z,t)\leqq 0{,}5\,\varrho_s(z,t)$ В соответствии с формулой Абэ:

$\varrho_s(z,t)=0{,}1854\cdot 10^{0{,}00545\,(l(t)-z)}$

$\lambda_s(z,t)=10^{-3}\,(6\varrho^4+1{,}9\varrho+0{,}05)$ формула Янсона [7]. Об'емная темпоемкость мерзлой почвы $c_1=c_{yg}\cdot\varrho_1$, где

$$c_{yg}=\frac{1}{m}(c_r m_r+c_B m_B+c_\Lambda m_\Lambda)=\frac{1}{w+1}(c_r+c_B w_1+c_\Lambda w_\Lambda)$$

(Здесь m, m_r, m_B, m_Λ — масса почвы, сухой почвы, воды и льда соответственно; c_r, c_B, c_Λ — удельная теплоемкость сухой почвы, воды, льда). Но $\varrho_1=(1+w)$. Поэтому $c_1=\gamma\,(c_r+c_B w_1+c_\Lambda w_\Lambda)$ Поскольку $c_r\cong 0{,}2$ кал/г град, $c_B=1$ кал/г град, $c_\Lambda=0{,}5$ кал/град, то $c_1=\gamma\,(0{,}2+w_1+0{,}5\,w_\Lambda)$. Для талой почвы $c_2=\gamma(0{,}2+w_2)$ коэффициенты $\lambda_j=\lambda_j(w)$, кривая $w_1=w_1(T)$ и $D_j=D_j(w)$ заимствовались из различных источников [4], [8], [9]. Величина δ^B_{Tj} бралась средней и равной $4\cdot 10^{-2}$ 1/°C [9], ибо найти ее зависимость от w для почв района ст. Белогорка не удалось. Что касается величины δ^T_{Bj}, то оценки, имеющиеся в литературе [9], позволяют принять ее в нашем

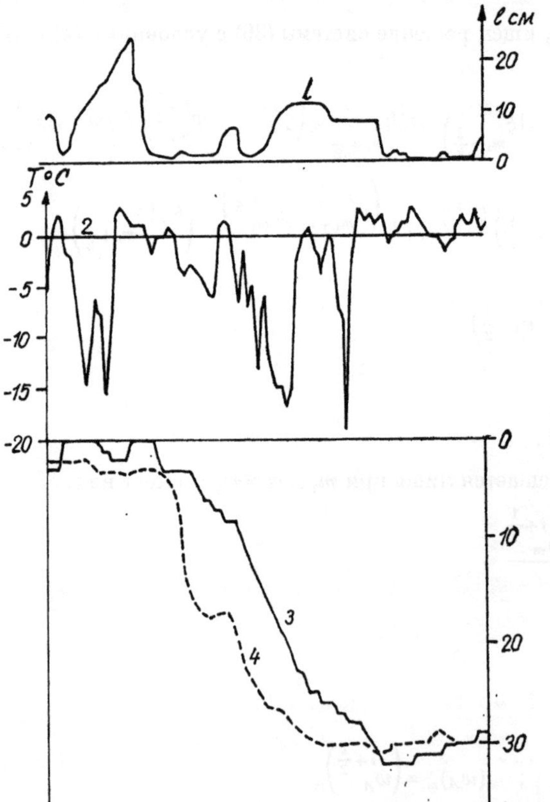

Рис. 1. Налюдения и расчет по данным метеостэнции Белогорка.
1 — высота снежного покрова
2 — среднесуточная температура воздуха
3 — наблюденная глубина промерзания в почве
4 — рассчитанная глубина промерзания в почве.

случае равной 0,01 °С. Поскольку точными данными мы не располагали, то расчеты проводились в диапазоне $0{,}001 < \delta^T_{Bj} \leqq 0{,}1$ °С. Отметим, что на окончательных результатах эти изменения практически не отражались. Это подтверждается данными всей серии расчетов. Таким образом, можно сделать вывод, что по крайней мере, в обычных естественных условиях влиянием массопереноса на температурный режим можно пренебречь.

На рис. 1 дан пример расчета по данным метеостанции Белогорка Ленинградской области за период с 31/XII-1974 г. по 27/III-1975 г. Начальная температура $T^\circ = -6{,}2$ °С, $T_H = 2°$, $W_H = 14\%$, $H = 100$ см, $\Delta t = 1$ сутки. Из рис. 2 следует, что миграция действительно имеет место. Совпадение рассчитанных и фактических наблюденных величин представляется вполне удовлетворительным.

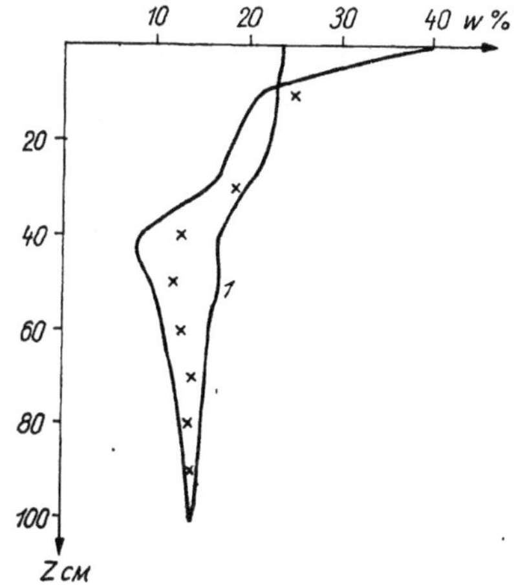

Рис. 2. Профили влажности в почве (м/с Белогорка)
1 — начальный 31/XII-1974 г.
2 — расчетный 27/III-1975 г.
x — наблюденные значения влажности 27/III-1975 г.

Литература

[1] *Лыков, А. Б., Ю. А. Михайлов.* Теория тепло- и массопереноса. Госэнергоиздат, М.-Л., 1963, 535 с.
[2] *де Гроот, С., П. Мазур.* Неравновесная термодинамика. «Мир», М., 1965, 456 с.
[3] *Дьярматти, И.* Неравновесная термодинамика. «Мир», М., 1974,
[4] *Глобус, А. М.* Экспериментальная гидрофизика почв. Гидрометеоиздат, Л., 1969, 355 с.
[5] *Марчук, Г. И.* Методы расчета ядерных реакторов. Госатомиздат, М., 1961.
[6] *Годунов, С. К., В. С. Рябенький.* Введение в теорию разностных схем. физматгиз, М., 1962, 340 с.
[7] *Кузьмин, П. П.* Физические свойства снежного покрова. Гидрометеоиздат, Л., 1967, 179 с.
[8] *Чудновский, А. Ф.* Теплофизические характеристики дисперсных материалов. ГИФИЛ, М., 1962, 456 с.
[9] *Нерпин, С. В., А. Ф. Чудновский.* Физика почвы. «Наука», М., 1967, 583 с.

Адреса авторов: В. Н. Лыкосов, 630090, Новосибирск-90, Вычислительный центр Сибирского отделения АН СССР, отделение физики атмосферы и океана, СССР;
П. Э. Палагин, 195196, Ленинград К 196, Малоохтенский проспект, Ленинградский Гидрометеорологический Институт, СССР

Индексы в тексте отвечают:

„s" — снегу; $j=1$ — мерзлой и $j=2$ — талой почве; $i=1, 2, 3, 4, 5$ — i -му компоненту.

Основные обозначения

t — время; z — вертикальная координата; T — температура; $T_B(t)$ — температура на уровне метеобудки.
$c(z, t)$, $c_p(z, t)$ — об'емная и удельная изобарная теплоемкость; $\lambda(z, t)$ — коэффициент теплопроводности;
α_i — масса i — компонента в единице об'ема, а v_i — скорость его движения относительно почвы; v — скорость движения центра масс; ϱ — плотность влажной почвы; ϱ_i — плотность i — компонента; $\gamma \equiv \alpha_2$ — об'емный вес сухой почвы;
w_B, w_A — весовая влажность воды и льда (количество воды, приходящееся на единицу веса сухой почвы);
I — источник вещества;
J_u, J_q, J — плотность молекулярного потока внутренней энергии, тепла и массы;
g — ускорение силы тяжести;
p — гидродинамическое давление;
h — удельная, h_i — парициальная удельная энтальпия;
L — теплота плавления льда;
u — удельная внутренняя энергия;
σ — источник энтропии;
δ_{ij} — символ Кронекера;
$l(t)$ — толщина снега;
$h(t)$ — положение нулевой изотермы;
H — уровень затухания сезонных колебаний температуры;
T_H — климатологическое значение температуры на уровне H.

551.543
551.509.313
551.507.22

Die operative numerische Analyse des Feldes der 3-stündigen Drucktendenz unter besonderer Berücksichtigung der Korrektur der Tendenz bei fahrenden Schiffen

Von **K. H. Stahnke-Jungheim** und **Ch. Vollbrecht**

Mit 1 Abbildung

Zusammenfassung: Das von *Belousov* u. a. [1, 2] entwickelte Schema zur objektiven Analyse des Drucktendenzfeldes wird durch die Korrektur der Drucktendenz bei fahrenden Schiffen verbessert. Diese Korrektur ist besonders bei hohen Schiffsgeschwindigkeiten und bei schnellen Veränderungen der Druckgebilde erforderlich. Das operativ eingesetzte Verfahren enthält neben einer erweiterten horizontalen Kontrolle Maßnahmen, die eine Information des Synoptikers (Klartext auf Plotterkarte) über extreme Beobachtungswerte bzw. die Beobachtungen isoliert liegender Stationen und deren horizontale Kontrolle ermöglichen.

Summary: The scheme developed by *Belousov* et al. [1, 2] for an objective analysis of the pressure tendency field is improved by a correction applied to the pressure tendency values as reported by moving ships. Such correction is required particularly for fast ship movements and/or with rapidly changing pressure patterns.

The routine method includes, apart from an extended horizontal control, some means to supply the forecaster with information (plotted charts with supplementary clear language information) on extreme observational data or observations made by remote stations and their horizontal control.

Резюме: Оперативная схема обьективного анализа тенденции давления, развитая *Белоусовым* [1, 2], усовершенствуется коррекцией тенденции давления для движущихся кораблей. Такая коррекция особенно нужна при высокой скорости кораблей и резких изменений барических образований. Оперативная схема содержит кроме расширенного горизонтального контроля специальные меры, которые доставляют синоптику специальную информацию (текст на карте), о сводках с экстремальными значениями, о сводках с изолированых станции и результате горизонтального контроля.

1. Einleitung

Eine objektive Analyse des Feldes der 3-stündigen Tendenz des Bodendruckes ist für den operativ tätigen Meteorologen als Arbeitsunterlage sowie als zusätzliche Information für die numerische Wetterprognose von Bedeutung (z. B. *Mercalov* [7]). Für den Synoptiker bedeutet die automatisch gezeichnete Analysekarte nicht nur eine erhebliche Zeiteinsparung, sondern wegen der Verwendung aller bis zu einem Redaktionsschluß eingegangenen Meldungen besonders über dem Atlantik eine Verbesserung der Qualität gegenüber der subjektiven Analyse. Die Verwendung aller eingegangenen Meldungen und die günstige Wahl des Redaktionsschlusses ist wegen des geringeren Korrelationsradius der Autocovarianzfunktion des Drucktendenzfeldes im Vergleich zum Feld des Bodendruckes von erheblicher Bedeutung. Fehlende Meldungen wirken sich bei der horizontalen Kontrolle und bei der Gitterpunktsinterpolation besonders über Seegebieten entscheidend aus.

Das Anbringen einer Korrektur an der beobachteten Drucktendenz fahrender Schiffe ist bei der heutigen Geschwindigkeit der Schiffe unbedingt erforderlich. Beispielsweise wird ein in ein stationäres Tief hineinfahrendes Schiff einen Druckfall melden. Die Zeit, die der Synoptiker für die Durchführung der Korrektur nach einer Tabelle ([6] S. 38, s. auch [8]) benötigt, ist erheblich, auch wenn er nur die Drucktendenzen eines Teils der 80–150 Schiffe verwendet, deren Beobachtungen im Analysegebiet (s. [5]) bis zu einem Redaktionsschluß von $T+140$ min vorliegen. Das im folgenden beschriebene Verfahren zur Analyse des Feldes der Drucktendenz wurde ausgehend von dem von *Belousov* u. a. [1, 2] vorgeschlagenen Schema unter Verwendung des von *Kluge*, *Stahnke-Jungheim* und *Vollbrecht* [5] entwickelten Modells zur subsynoptischen Einflächenanalyse erarbeitet. Es gliedert sich in 5 Teile:

– Berechnung der Korrektur der Drucktendenz bei fahrenden Schiffen
– Geographische Sortierung der Beobachtungen
– Horizontale Kontrolle der Beobachtungen
– Berechnung der Drucktendenzwerte für die Gitterpunkte eines regulären Netzes (Gitterpunktsinterpolation)
– Bereitstellen von zusätzlicher Information für den Synoptiker

2. Die Korrektur der Drucktendenz bei fahrenden Schiffen

Nachdem alle bis zum jeweiligen Redaktionsschluß (z. Z. $T+140$ min) eingetroffenen und der EDVA angebotenen Bodenwettermeldungen eines vorgegebenen Gebietes von der Primären Datenbearbeitung (ERKENNUNG) erkannt, decodiert, logisch und z. T. meteorologisch geprüft wurden und in einem bestimmten Format auf einem externen Speichermedium (Magnetband, Trommel, Platte) bereitgestellt sind (vgl. [5]), wird für alle Handelsschiffe die Korrektur der Drucktendenz auf der Grundlage der von *Randall* [9] angegebenen Beziehung vorgenommen. Für ein Schiff, das auf seinem Kurs Isobaren kreuzt, gilt (vgl. Abb. 1):

$$(1)\qquad \frac{\partial p}{\partial t}=\frac{dp}{dt}+F\left(\mathfrak{v}_s,\,\mathfrak{v},\,\frac{\partial p}{\partial n}\right),\ \left[\frac{\text{mbar}}{3\,\text{h}}\right],$$

wobei $\frac{\partial p}{\partial t}$ die tatsächliche 3-stündige Druckänderung, $\frac{dp}{dt}$ die gemessene 3-stündige Druckänderung und F die Korrektur der Drucktendenz in Abhängigkeit von der Schiffsbewegung $\mathfrak{v}_s$ (Kurs, Geschwindigkeit), dem Windvektor $\mathfrak{v}$ (Richtung, Geschwindigkeit) und dem Druckgradienten $\frac{\partial p}{\partial n}$ ist. Für die Korrektur F läßt sich leicht folgende Beziehung finden (s. Abb. 1):

$$F = -3\,\frac{\partial p}{\partial n}\, v_s \sin\alpha, \left[\frac{mbar}{3h}\right], \tag{2}$$

mit $\frac{\partial p}{\partial n}$ dem Druckgradienten des Isobarenfeldes (mbar/km), v_s der mittleren Schiffsgeschwindigkeit der letzten 3 Stunden (km/h) und α der Abweichung des mittleren Kurses der letzten 3 Stunden vom Gradientwind. Es gilt

$$\alpha \doteq d_s - (d+\gamma) = d - d_s - \gamma\,, \tag{3}$$

d_s ist der mittlere Schiffskurs, d die Windrichtung und γ die Abweichung des Gradientwindes vom wahren (beobachteten) Wind. Eine gewisse Ungenauigkeit beim Bestimmen der Größe α ist unvermeidlich, da nur eine Angabe der Genauigkeit des Kurses auf 45° und der Windrichtung auf 10° möglich ist. Dazu kommt noch das Abschätzen der Gradientwindabweichung γ.

Ähnlich wie mit der Bestimmung des Kurses verhält es sich mit der Berechnung der Schiffsgeschwindigkeit v_s, auch hier erlauben die Codezahlen nur eine ungefähre Angabe der tatsächlichen Geschwindigkeit (auf ganze Zahlen gerundete Werte, s. Tab. 1).

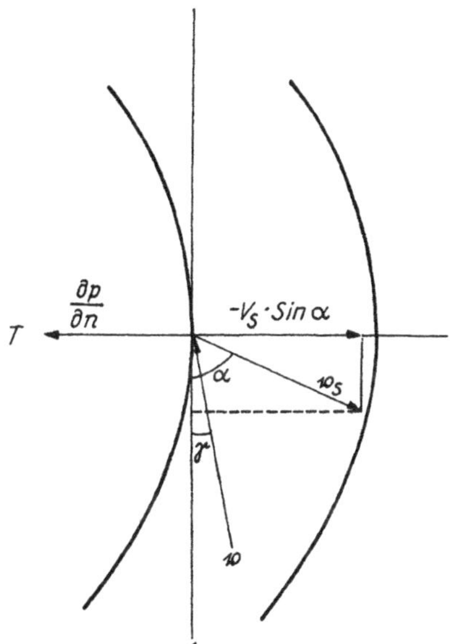

Abb. 1. Bewegung eines Schiffes im Isobarenfeld
$\mathfrak{v}_s$ = Schiffsbewegung
$\mathfrak{v}$ = Windvektor
$\frac{\partial p}{\partial n}$ = Druckgradient
t = Tangente zum Isobarenfeld

Tabelle 1

Zuordnung des Codes der Schiffsgeschwindigkeit zur Geschwindigkeit

Code	0	1	2	3	4	5	6	7	8	9
Geschwindigkeit in Knoten	0	1–3	4–6	7–9	10–12	13–15	16–18	19–21	22–24	≧25

Beim Festlegen der Werte für d_s und v_s wurde jeweils die Klassenmitte der Richtungs- bzw. Geschwindigkeitsbereiche verwendet.

Zum Bestimmen der Korrektur F sind noch die Größen $\frac{\partial p}{\partial n}$ und γ anzugeben. Den Druckgradienten $\frac{\partial p}{\partial n}$ erhalten wir aus der Gradientwindgleichung

$$\frac{\partial p}{\partial n} = -l v_g \varrho - \frac{v_g^2 \varrho}{r}\,. \tag{4}$$

Hierin bedeuten l den Coriolisparameter, v_g die Gradientwindgeschwindigkeit, ϱ die Luftdichte und r den Krümmungsradius der Bahnlinie eines Luftteilchens gleicher Position wie das Schiff. Für zyklonale Bahnlinien ist r positiv, für antizyklonale negativ.

Mit (2)–(4) wird aus (1):

$$\frac{\partial p}{\partial t}=\frac{dp}{dt}+3\left(lv_g\varrho+\frac{v_g^2\varrho}{r}\right)v_s\sin(d_s-d-\gamma)\,. \tag{5}$$

Für unsere Zwecke genügt es, für die Luftdichte eine Konstante zu benutzen und die Gradientwindgeschwindigkeit v_g in Abhängigkeit von der ebenfalls konstant zu wählenden Gradientwindabweichung γ proportional der Windgeschwindigkeit v zu wählen:

$$v_g=av\cdot \tag{6}$$

Für die Luftdichte ϱ verwenden wir den Wert für 1 km über *NN* und 8,5 °C:

$$\varrho=1{,}112\cdot10^{-3}\,\frac{g}{cm^3}\,.$$

Für a und γ benutzen *Olson* [8], *Linke* [6] und *Randall* [9] in ihren Umrechnungstabellen bzw. Graphiken folgende Werte (s. Tab. 2):

Tabelle 2

Zuordnung der Gradientwindabweichung zum Verhältnis von Gradientwindgeschwindigkeit und Windgeschwindigkeit

	γ	a
Olson	20°	$\frac{5}{4}=1.25$
Linke		
Randall	15°	$\frac{3}{2}=1.50$

Damit ist in Gleichung (5) nur noch der Krümmungsradius r unbekannt. Er könnte nur aus einem bereits vorliegenden Bodendruckfeld ermittelt werden. *Olson* [8] und *Linke* [6] vernachlässigen in ihren für die Praxis entwickelten Umrechnungstabellen und Graphiken den Summanden $v_g^2\,\varrho/r$, da er im allgemeinen klein gegenüber der Größe $lv_g\varrho$ ist.

Unter den genannten Annahmen und der Beziehung

$$l=1{,}458\cdot10^{-4}\sin\varphi\left[\frac{1}{s}\right] \tag{7}$$

für den Coriolisparameter ergibt sich aus (5) für die tatsächliche 3-stündige Druckänderung [mbar/3 h]:

$$\frac{\partial p}{\partial t}\approx\frac{dp}{dt}+4{,}863\cdot10^{-3}av\cdot v_s\cdot\sin\varphi\cdot\sin(d_s-d-\gamma) \tag{8}$$

mit folgenden Maßeinheiten:

$\frac{dp}{dt}$: gemeldete Drucktendenz [mbar/3 h]
a : Verhältnis der Gradientwindgeschwindigkeit zur Windgeschwindigkeit (s. Tab. 2)
v : Windgeschwindigkeit $\left[\frac{m}{s}\right]$
v_s : mittlere Schiffsgeschwindigkeit $\left[\frac{km}{h}\right]$
φ : geographische Breite [Grad]
d : mittlerer Schiffskurs [Grad]
d_s : Windrichtung [Grad]
γ : Gradientwindabweichung [Grad] (s. Tab. 2).

Die Berechnung der Korrektur der Drucktendenz wird nun im Programmsystem der numerischen Analyse folgendermaßen organisiert. Jede vom Programmpaket ERKENNUNG auf einem externen Speicherbereich in einem bestimmten Format abgelieferte Bodenwettermeldung wird geprüft, ob sie von einem Handelsschiff stammt, wenn ja, so müssen folgende Kriterien erfüllt sein, damit die Korrektur errechnet werden kann bzw. deren Bestimmung sinnvoll ist:

- $\frac{dp}{dt}$, v, d, v_s, d_s und φ müssen gemeldet sein und nicht als „unzuverlässig" gekennzeichnet sein (unzulässiger Code, meteorologisch unsinnig).
- „Windstille" oder „umlaufende Winde" darf nicht vorliegen
- für die Windgeschwindigkeit v, die Schiffsgeschwindigkeit v_s und die Abweichung α des Kurses vom Gradientwind müssen folgende vom Synoptiker empfohlene Grenzen überschritten sein:

$$\begin{aligned}&v>3\ m/s\\&v_s\geqq 8\ sm/h\ (\text{Code 3 für } v_s)\\&|\alpha|=|d_s-d-\gamma|>25°.\end{aligned} \tag{9}$$

Für Meldungen, die diesen Kriterien genügen, wird die Korrektur gemäß Gleichung (8) berechnet. Für alle Meldungen wird ein reduziertes Format, das neben den geographischen Koordinaten den Stationsindex (bei Landstationen), die gemeldete oder korrigierte Drucktendenz und ein Wort mit organisatorischen und Unzuverlässigkeitskennungen enthält, als Ausgangsinformation für die objektive Analyse bereitgestellt.

3. Die numerischen Analyse des Drucktendenzfeldes

Die weitere Verarbeitung der Meldung erfolgt analog dem in [5] vorgestellten Verfahren. Die Meldungen werden geographisch sortiert nach Umrechnung der geographischen Koordinaten in kartesische Koordinaten des Quadratnetzgebietes. Danach wird die horizontale Kontrolle der Drucktendenzmeldungen durchgeführt (vgl. [1, 2, 5]):

(10) $$\left|\left(\frac{dp}{dt}\right)_k - \left(\frac{dp}{dt}\right)_k^*\right| \leqq \sigma\sqrt{\varepsilon+\eta}\,,$$

$\left(\frac{dp}{dt}\right)_k$ ist die Drucktendenz der k-ten zu kontrollierenden Station, $\left(\frac{dp}{dt}\right)_k^*$ die durch Interpolation aus 4 umliegenden Beobachtungswerten gewonnene Schätzgröße, ε der theoretische Interpolationsfehler $\eta = 0{,}02$ der relative Beobachtungsfehler und $\sigma = 6{,}4$ mbar eine konstante Dispersion. Für $\left(\frac{dp}{dt}\right)_k^*$ gilt:

(11) $$\left(\frac{dp}{dt}\right)_k^* = \sum_{i=1}^{4}\left(\frac{dp}{dt}\right)_i b_i\,.$$

Der Grundfeldwert für jede Beobachtung bzw. für jeden Gitterpunkt ist gleich Null, b_i sind die Gewichte. Als Korrelationsfunktion wird für die horizontale Kontrolle und Gitterpunktsinterpolation die Beziehung (s. [5]) verwendet:

(12) $$K(r_{ij}) = \left(1+\frac{r_{ij}}{L}\right) e^{-r_{ij}/L}\,,$$

mit dem Korrelationsradius $L = 0{,}3$ (für das Bodendruckfeld $L = 0{,}8$) und r_{ij} dem Abstand der i-ten von der j-ten Station.

Die Kontrolle wird bei Fehlerverdacht (Gleichung (10) ist nichterfüllt) mit einer 5. und 6. Meldung unter Verwerten der von der ERKENNUNG gegebenen Unzuverlässigkeitskennungen durchgeführt (vgl. [5]).

Im Anschluß an die Kontrolle werden mit der Methode der optimalen Interpolation die Werte der Drucktendenz an den Punkten eines regulären Gitters nach (11) berechnet. Das Analysegebiet ist mit dem in [5] genannten Gebiet identisch. Der Gitterpunktsabstand beträgt 150 km. Größere Abstände führen wegen der geringeren Korrelation (s. Abschn. 1, vgl. die oben genannten Werte für L) im Drucktendenzfeld zu einer das Ergebnis verfälschenden Glättung. Da trotzdem das analysierte Feld noch eine zu starke Glättung besitzt und auch die Korrelationsfunktion (12) die statistische Struktur des Drucktendenzfeldes nur näherungsweise erfaßt, wird eine Verbesserung der Analyse durch die inverse Filterung nach *Striževskij* ([10], s. auch [3], [4]) vorgenommen:

(13) $$W_0 = 1{,}08 w_0 + 0{,}08 \sum_{i=1}^{4} w_i - 0{,}1 \sum_{i=5}^{8} w_i\,.$$

In Gleichung (13) bedeuten: W_0 der verbesserte Gitterpunktswert, w_0 der ursprüngliche Wert, w_1 bis w_4 die im Gitterpunktsabstand und w_5 bis w_8 die im Gitterpunktsabstand mal $\sqrt{2}$ von w_0 liegenden ursprünglichen Gitterpunkte.

4. Ergebnisse der operativen Berechnung der Drucktendenzanalyse Zusätzliche Information für den Synoptiker

Eine numerische Drucktendenzanalyse des vorgestellten Schemas wird operativ zum 00^{00}- und 12^{00}-Termin in der Zentralen Wetterdienststelle des Meteorologischen Dienstes der DDR berechnet und als Plotterkarte dem Synoptiker zur Verfügung gestellt. Der Redaktionsschluß für die eintreffenden Meldungen beträgt $T+140$ min. Zu diesem Zeitpunkt liegen im betreffenden Gebiet etwa 1000–1300 Meldungen vor (s. [5]).

Die Rechenzeit beträgt für die Korrektur der Drucktendenz, die Sortierung, die Kontrolle, die Gitterpunktsinterpolation für ein 45×53 Analysegitter (150 km Gitterweite) 4 min und die Berechnung der Plotterkarte auf der EDVA BESM-6 2 min, die Zeichnung selbst in Abhängigkeit von der Wetterlage 6 bis 8 min. Die Karte, die in der Technologie des Redaktionsschlusses von $T+140$ min erarbeitet wird, steht dem Synoptiker etwa zum Zeitpunkt $T+180$ min in dreifarbiger Ausführung zur Verfügung (blau: positive Isallobaren, grün: Null-Isallobare, rot: negative Isallobaren, Isolinienabstand 1 mbar).

Die operative Arbeit brachte hinsichtlich der Berechnung der Korrektur der Drucktendenz bei fahrenden Schiffen folgende Ergebnisse:

– die Anwendung der beiden in Tab. 2 angegebenen Wertepaare für die Gradientwindabweichung und das Verhältnis der Gradientwindgeschwindigkeit zur Windgeschwindigkeit ergab für die Korrektur nur unwesentliche Unterschiede (operativ wird $\gamma = 15°$ und $a = 1{,}5$ verwendet)

- die vom Synoptiker angegebenen Grenzwerte für die Schiffs- und Windgeschwindigkeit und die Abweichung des Kurses vom Gradientwind (s. Beziehung (9)), bei deren Überschreitung die Berechnung der Korrektur erst sinnvoll ist, wurden durch die Praxis bestätigt
- die Drucktendenz von 25–35% der etwa 80–150 Schiffsmeldungen, die bis zum Redaktionsschluß vorliegen, wird korrigiert
- die Korrektur beträgt in Abhängigkeit vom Kurs und der meteorologischen Situation 0 bis ±4 mbar.

Durch die horizontale Kontrolle werden etwa 1,5 bis 2,5% der Meldungen verworfen.

Die horizontale Kontrolle der Drucktendenz ist besonders bei extremen, aber synoptisch richtigen Meldungen und/oder isoliert liegenden Stationen wegen der geringen Korrelation weit problematischer als die horizontale Kontrolle des Bodendrucks. Wir können diese Schwierigkeit nur durch eine weitere Verbesserung des in [5] beschriebenen Kontrollalgorithmus und durch zusätzliche Information des Synoptikers in Klartext auf der Plotterkarte als eine bescheidene Form der Mensch-Maschine-Kombination überwinden.

Wegen der Gefahr des Verwerfens der Meldungen isoliert liegender Stationen durch die horizontale Kontrolle werden in Klartext für alle verworfenen isoliert liegenden Stationen (i. a. nur Schiffe) mit einem Betrag der Drucktendenz größer gleich 3 mbar die Koordinaten, der Stationsindex (Landstation) und die beobachtete bzw. korrigierte Drucktendenz auf den Rand der Plotterkarte geschrieben. Als Maß für die Isoliertheit gilt $\sigma\sqrt{\varepsilon} > 2{,}0$ bzw. $\varepsilon > 0{,}096$, das entspricht einem mittleren Radius der 4 umliegenden Stationen von 150 km.

Weiterhin werden als zusätzliche Information, alle Meldungen mit einem Betrag der Drucktendenz größer gleich 9,9 mbar ausgeschrieben, da einmal international oft die Zusatzgruppen des FM 11, die u. a. auch die Drucktendenz größer 9,9 mbar enthalten, nicht verbreitet werden und zum anderen Werte größer 10 mbar eben extreme Werte sind, die durch die horizontale Kontrolle durchaus verworfen werden können.

Die Drucktendenzmeldungen der Wetterschiffe werden wegen der Gefahr des Verwerfens nicht horizontal kontrolliert. Bei Vorliegen einer Unzuverlässigkeitskennung für die Tendenz durch die ERKENNUNG wird diese Beobachtung nicht verwendet.

Mit diesen bisher erwähnten Maßnahmen ist aber das Problem der Kontrolle extremer aber synoptisch richtiger Werte noch nicht gelöst. Nicht nur die Werte isoliert liegender Stationen werden durch die horizontale Kontrolle verworfen, sondern auch nicht isolierte Stationen wegen der geringen Korrelation und der besonders in kräftigen Steig- und Fallgebieten durchaus beträchtlichen Unterschiede bis mehr als 4 mbar zwischen benachbarten Stationen.

Der in [5] beschriebene Kontrollalgorithmus, basierend auf dem Verfahren der optimalen Interpolation, muß hier versagen. Deshalb schlagen wir vor, bevor eine Meldung nach der Kontrolle mit 4 Stationen und einer 5. und 6. Station verworfen werden soll, wird untersucht, ob vielleicht ein extremer Wert in einem ausgeprägtem Fall- oder Steigegebiet vorliegt. Als Kriterium dafür gilt: Der Wert der zu prüfenden Meldung ist betragsmäßig größer oder gleich 4 mbar und der maximale interpolierte Wert ist bei gleichem Vorzeichen wie die zu kontrollierende Meldung betragmäßigs größer oder gleich 3 mbar.

Ist dieses Kriterium erfüllt, so wird geprüft:

$$\left|\left(\frac{dp}{dt}\right)_k - \left(\frac{dp}{dt}\right)_k^*\right| \leqq 2\sigma\sqrt{\varepsilon+\eta}\,. \tag{14}$$

Ist die Gleichung nicht erfüllt, so wird der zu kontrollierende extreme Wert endgültig verworfen. Alle auf diese Weise verworfenen Meldungen werden ausgeschrieben und als solche gekennzeichnet, wie auch die oben erwähnten bei Verwerfen markiert werden. 80 Meldungen können auf die Plotterkarte geschrieben werden.

Die Prüfung extremer Drucktendenzen kann zusätzlich durch Betrachten der Druckwerte und ihrer Umgebung erfolgen. Das wäre dann ein erster Schritt zur komplexen Kontrolle der meteorologischen Bodeninformation (vgl. [5]).

Gegenwärtig wird als eine weitere Verbesserung der Klartextinformation eine Variante realisiert, die nicht nur das Ausschreiben der Tendenzen sondern der gesamten Meldung (6 Hauptgruppen) auf die Plotterkarte erlaubt.

Für die Zukunft sind folgende Verbesserungen der Korrektur und des Analyseschemas vorgesehen. Bei der Korrektur der Drucktendenz erscheint es zweckmäßig, den beobachteten Wind durch einen aus dem Druckfeld errechneten (vgl. [5], Verfahren der Windkontrolle) zu ersetzen, da die Bestimmung des Windes wegen der hohen Schiffsgeschwindigkeit und des relativ ungeschulten Personals immer mit einer gewissen Ungenauigkeit versehen ist. Dieses Ersetzen verlangt die Verwendung der Bodendruckwerte in der Drucktendenzanalyse oder bereits das Vorliegen der Bodendruckanalyse.

In einer weiteren Arbeit soll über den Aufbau eines operativen Analysesystems für die meteorologische Bodeninformation (auch Temperatur- und Taupunktanalysen) als Anfangswertlieferant für die Numerische Wetterprognose und Erzeuger von Arbeitsunterlagen für den Synoptiker berichtet werden.

Die Autoren möchten Herrn Dipl.-Met. *Eckner* für seine Hinweise beim Erarbeiten des Analyseschemas besonders hinsichtlich der Korrektur der Drucktendenz sowie Herrn *Dr. Belousov* und Herrn *Dr. Kluge* für ihre Anregungen zur Weiterführung der Arbeit danken.

Ebenso danken sie Herrn Dipl.-Met. *Lambacher* für die Realisierung der Änderungen und Ergänzungen im Modell der subsynoptischen Einflächenanalyse (s. [5]).

Literatur

[1] *Bagrov, A. N., S. L. Belousov, A. K. Artanova, A. G. Tarnopot'skij,* Operativnaja schema ob'ektivnogo analiza prizemnych meteorologičeskich polej na severnom polušarii dlja prjamougolnoj setki s peremennym šagom. *Trudy GMC,* 123 (1974), S. 37–47

[2] *Bagrov, A. N., S. L. Belousov, A. G. Tarnopol'skij,* Schema ob'ektivnogo analiza polej davlenija baričeskoj tendencii, temperatury i točki rosy po dannym seti sinoptičeskich stancij. *Trudy GMC,* 91 (1972), S. 58–76

[3] *Belousov, S. L., L. N. Striževskij,* Sposob učeta melkomasštabnych sostavljajuščich pri ob'ektivnom analize polej baričeskoj tendencii i prizemnogo davlenija. *Meteor. i Gidro.,* 1969, №. 5, S. 13–20

[4] *Belousov, S. L., L. N. Striževskij,* Operativnaja schema ob'ektivnogo analiza prizemnogo davlenija i baričeskich tendenzij dlja učaščennoj setki *Trudy GMC,* 54 (1970), S. 3–10

[5] *Kluge, J., K. H. Stahnke-Jungheim, Ch. Vollbrecht,* Die Verwendung von Winddaten bei der operativen numerischen Analyse des Bodendruckfeldes. Z. Meteor., *27* (1977), Heft 3, S. 147–155

[6] *Linke, F., F. Baur,* Meteorologisches Taschenbuch, I. Band, Leipzig 1962

[7] *Mercalov, A. N.,* K prognozu baričeskogo polja v nižnem sloe atmosfery Trudy GMC, 57 (1969), S. 3–20

[8] *Olson, B. E.,* Correction of Ship Pressure Tendencies for Cross-Isobar Movement of Vessel. Bull. Amer. Met. Soc., 27 (1946), S. 314–318

[9] *Randall, D. L.,* A Method for the Correction of 3-hour Pressure Tendencies in Ship Reports. Bull. Amer. Met. Soc., 22 (1941), S. 14–16

[10] *Striževskij, L. N.,* Ob optimal'noj filtracii meteorologičeskich polej Trudy CIP, 146 (1965)

Anschrift: Dr. K. H. Stahnke-Jungheim, Dipl.-Math. Ch. Vollbrecht, Meteorologischer Dienst der DDR, Zentrale Wetterdienststelle, DDR-15 Potsdam, Michendorfer Chaussee 23

551.509.313

Lineare Untersuchung eines speziellen semi-impliziten Zeitintegrationsverfahrens für ein Mehrflächenmodell*)

Von **H.-J. Herzog**

Mit 2 Abbildungen

Zusammenfassung: Ein baroklines semi-implizites Modell wird einer linearen Untersuchung unterzogen. Das Zeitschema desselben weist folgende Eigenschaften auf. Es ist selektiv dämpfend, insofern nur der Gravitationswellenanteil damit beeinflußt wird. Der synoptisch relevante Lösungsanteil wird in seiner Amplitude praktisch nicht modifiziert, wenn ein geeigneter lateraler Diffusionskoeffizient eingeführt wird. Die Phasenfehler sind vernachlässigbar. Das Zeitschema verwendet aus rechenökonomisch wünschenswerten Gründen nur zwei Zeitniveaus und erzeugt deshalb keinen rechenbedingten Lärm.

Die lineare Analyse bedient sich einer Methode, die die baroklinen Ausgangsgleichungen auf einen von *Kurihara* (1965) untersuchten Gleichungstyp zurückführt.

Summary: A linear investigation is made of a baroclinic semi-implicit model. The following properties of this time scheme are outlined: It is a selective damping scheme. The amplitudes of the Rossby solution part are maintained, if a determined lateral diffusion coefficient is considered. The amplitudes of the gravitational solution part are damped. The greater the phase speed of the gravity modes the greater the damping effect. The phase errors in connection with this time scheme are negligible. The assumed scheme is a two-time level scheme only and therefore saves computing capacity. The computational mode does not occur. The given linear analysis applies a method from which it is possible to reduce the initially given linear baroclinic equations to much more simple wave-type equations already studied by *Kurihara* (1965).

1. Einführung

Entscheidet man sich bei der numerischen Integration der hydrothermodynamischen Gleichungen für die Klasse der primitiven, hydrostatischen Gleichungen, so zwingt der zusätzlich zu dem synoptisch wesentlichen Lösungsanteil enthaltene Lösungsanteil an Trägheits- und Gravitationswellen zu besonderer Vorsicht bei der numerischen Rechnung und, dem vorausgehend, zu ganz spezifischen Überlegungen hinsichtlich der Modellierung. Dabei sind auf jeden Fall wenigstens die fünf von *Miyakoda* (1962) aufgezählten Forderungen an das konzipierte numerische Schema des Modells von nahezu gleichrangiger, sich gegenseitig bedingender Wichtigkeit. Erstens muß die Rechnung notwendigerweise stabil sein, zweitens soll die Genauigkeit der Lösung so hoch wie möglich sein, drittens ist die Einfachheit des numerischen Schemas wünschenswert, viertens ist bei vorgegebener Rechenmaschinenkapazität die damit verbundene Hauptspeicherbelastung zu beachten und fünftens soll der so aufgestellte Algorithmus rechnerisch schnell sein. In diesem Zusammenhang wird klar, daß die Erfüllung aller genannter Forderungen gleichermaßen ein wünschenswerter Idealzustand ist und tatsächlich die Beachtung eines Gesichtspunktes teilweise auf Kosten eines oder aller anderen gehen kann. Die Kunst besteht folglich bei der Modellierung darin, bei unbedingter Einhaltung von Punkt eins auf alle Fälle im Rahmen der primitiven Gleichungen ein möglichst ökonomisches Rechenschema bezüglich geringer Hauptspeicheranforderungen und großer Rechengeschwindigkeit zu wählen. Es eröffnet sich damit die grundsätzliche Chance, auf einer mittleren Rechenananlage mit den primitiven Gleichungen rechnen zu können. Wenn die verfügbare Rechenmaschine über relativ mehr Kapazität verfügt, so ist mit einem solchen Schema durch Verfeinerung der horizontalen und vertikalen Maschenweite, und was besonders wesentlich ist, durch Einführung subgitterskaliger Phänomene die Freiheit gewonnen, im Sinne von Punkt zwei die Genauigkeit des Modells entsprechend hochzutreiben.

Bekanntlich bieten zur Lösung der primitiven Gleichungen implizite und semi-implizite Zeitintegrationsverfahren eine gute Möglichkeit, die obengenannten ökonomischen Gesichtspunkte zu beachten (*Thompson* 1961, *Bortnikov* 1962, *Kurihara* 1965, *Marčuk* 1967, *Robért* 1968, *Kwizak* und *Robért* 1971 u. v. a.). Von besonderem Interesse sind die seit *Robért* 1968 vielerorts verwendeten semi-impliziten Integrationsverfahren. Die Idee beruht dabei in der nur impliziten Behandlung der für Gravitationswellen repräsentativen Terme, während der advektive Teil des Gleichungssystems explizit (leap-frog-Schema) gerechnet wird. Im Ergebnis wird der ökonomische Vorteil erreicht, mit Zeitschritten rechnen zu können, wie sie mit vorgegebener horizontaler Maschenweite des Rechengitters bei gefilterten Modellen üblich sind. Dabei hat die Erfahrung gezeigt, daß die über zehnmal größeren Zeitschritte gegenüber rein expliziten Schemata praktisch keine größeren zeitlichen Trunkationsfehler mit sich bringen und bei baroklinen Modellen eine bis zu 3,5-fach schnellere Rechengeschwindigkeit ergeben können. Ein solches Zeitschema birgt einerseits den Vorteil in sich, gefahrlos wesentlich größere Zeitschritte verwenden zu können, macht aber drei Zeitniveaus notwendig. Speichertechnisch folgt daraus, daß für jede Prognosevariable gleichzeitig zwei Felder zu aufeinanderfolgenden Zeitniveaus im Hauptspeicher für die Rechnung verfügbar sein müssen. Im Zusammenhang mit dem oben gesagten über die Bedeutung der Speicherökonomie wäre folglich ein semi-implizites Schema, welches nur zwei Zeitniveaus benutzt und folglich für jede Rechenvariable nur ein Feld im Hauptspeicher benötigt, wünschenswert.

Wir beabsichtigen hier, ein derartiges semi-implizites Zeitschema mit nur zwei Zeitniveaus zu untersuchen. Die Untersuchung wird anhand eines linearisierten, primitiven N-Flächenmodells ausgeführt und dabei das Verhal-

*) Herrn Prof. Dr. F. Bernhardt zum 80. Geburtstag gewidmet.

ten der Amplituden und Phasen für die meteorologisch advektiven Wellen als auch die Trägheits- und Gravitationswellen qualitativ bestimmt. Für ein spezielles 2-Flächen-Modell wird eine entsprechende quantitative Analyse mitgeteilt.

2. Das lineare N-Flächen-Modell

Wir wählen das folgende, linearisierte barokline Modell, an dem die beabsichtigten Untersuchungen über Amplituden- und Phasenverhalten für ein bestimmtes semi-implizites Zeitschema angestellt werden sollen. Die Methode unseres Vorgehens ist dabei nicht an das spezielle Zeitschema gebunden, sondern demonstriert bis zu einem gewissen Punkt auch, wie für ein baroklines N-Flächen-Modell jedes andere Zeitschema entsprechend untersucht werden kann. Das linearisierte System lautet

(1) $$\frac{\partial u}{\partial t}+U\frac{\partial u}{\partial x}-fv+\frac{\partial \Phi}{\partial x}=A\frac{\partial^2 u}{\partial x^2}$$

(2) $$\frac{\partial v}{\partial t}+U\frac{\partial v}{\partial x}+fu=A\frac{\partial^2 v}{\partial x^2}$$

(3) $$\frac{\partial}{\partial t}\left(\frac{\partial \Phi}{\partial p}\right)+U\frac{\partial}{\partial x}\left(\frac{\partial \Phi}{\partial p}\right)+\sigma\omega=A\frac{\partial^2}{\partial x^2}\left(\frac{\partial \Phi}{\partial p}\right)$$

(4) $$\frac{\partial u}{\partial x}+\frac{\partial \omega}{\partial p}=0$$

mit den Randbedingungen

(5) $\omega=0$ für $p_0=0$

und

(6) $w=0$ für $p_*=1000$ mbar.

Mit der unteren Randbedingung (6) sind externe Gravitationswellen gestattet. In der Bezeichnungsweise hier und im folgenden halten wir uns an die übliche Symbolik, um diesbezügliche Erläuterungen zu vermeiden. Die zonale Grundströmung U, die statische Stabilität σ, der Coriolisparameter f und der laterale Diffusionskoeffizient A seien Konstanten. Die in (1), (2), (3) zusätzlich angefügten Diffusionsterme sind in ihrer Bedeutung vorerst offen und könnten auch mit $A=0$ unberücksichtigt gedacht sein. In den folgenden Rechnungen führen wir diese zunächst jedoch rein formal mit. Für das weitere ist es zweckmäßig, die thermodynamische Gleichung (3) umzuformen, indem die Vertikalgeschwindigkeit ω mit Hilfe der Kontinuitätsgleichung (4) eliminiert wird:

(7) $$\frac{\partial}{\partial t}\left[-\frac{\partial}{\partial p}\left(\frac{1}{\sigma}\frac{\partial \Phi}{\partial p}\right)\right]+U\frac{\partial}{\partial x}\left[-\frac{\partial}{\partial p}\left(\frac{1}{\sigma}\frac{\partial \Phi}{\partial p}\right)\right]+\frac{\partial u}{\partial x}=A\frac{\partial^2}{\partial x^2}\left[-\frac{\partial}{\partial p}\left(\frac{1}{\sigma}\frac{\partial \Phi}{\partial p}\right)\right]$$

Für die Randbedingungen (5) und (6) folgt mit (3)

(8) $$\frac{\partial}{\partial t}\left(\frac{1}{\sigma}\frac{\partial \Phi}{\partial p}\right)+U\frac{\partial}{\partial x}\left(\frac{1}{\sigma}\frac{\partial \Phi}{\partial p}\right)=A\frac{\partial^2}{\partial x^2}\left(\frac{1}{\sigma}\frac{\partial \Phi}{\partial p}\right)\quad (\text{für } p_0=0)$$

bzw.

(9) $$\frac{\partial}{\partial t}\left(\frac{1}{\sigma}\frac{\partial \Phi}{\partial p}+\varrho\Phi\right)+U\frac{\partial}{\partial x}\left(\frac{1}{\sigma}\frac{\partial \Phi}{\partial p}+\varrho\Phi\right)=A\frac{\partial^2}{\partial x^2}\left(\frac{1}{\sigma}\frac{\partial \Phi}{\partial p}\right)\quad (\text{für } p=p_*),$$

so daß für weiteres das Gleichungssystem (1), (2), (7) mit den Randbedingungen (8) und (9) bereitgestellt ist. ϱ bedeutet hier eine Standarddichte. Schließlich kann die vertikale Diskretisierung am N-Flächen-Modell ausgeführt werden. Die Vertikalstruktur des N-Flächen-Modells ist in Abb. 1 wiedergegeben.

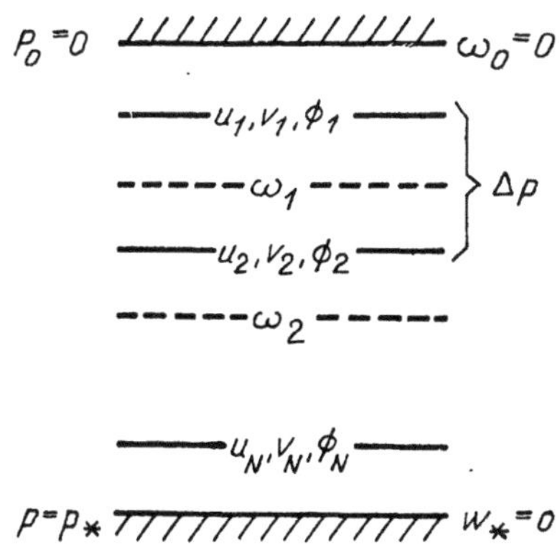

Abb. 1.

Werden durchweg zentrierte Differenzen angewendet – das gelingt nur nicht wegen der Randbedingung (9) – so ergibt sich mit den eingearbeiteten Randbedingungen (8) und (9) das Gleichungssystem

$$\frac{\partial \boldsymbol{u}}{\partial t}+U\frac{\partial \boldsymbol{u}}{\partial x}-f\boldsymbol{v}+\frac{\partial \boldsymbol{\Phi}}{\partial x}=A\frac{\partial^2 \boldsymbol{u}}{\partial x^2} \tag{10}$$

$$\frac{\partial \boldsymbol{v}}{\partial t}+U\frac{\partial \boldsymbol{v}}{\partial x}+f\boldsymbol{u}=A\frac{\partial^2 \boldsymbol{v}}{\partial x^2} \tag{11}$$

$$\boldsymbol{A}\left(\frac{\partial \boldsymbol{\Phi}}{\partial t}+U\frac{\partial \boldsymbol{\Phi}}{\partial x}-A\frac{\partial^2 \boldsymbol{\Phi}}{\partial x^2}\right)+\sigma(\Delta p)^2\frac{\partial \boldsymbol{u}}{\partial x}=0 \tag{12}$$

$\boldsymbol{u}$, $\boldsymbol{v}$, $\boldsymbol{\Phi}$ sind hier Vektoren, deren Komponenten die u, v, Φ auf den einzelnen Flächen bedeuten, also

$$\boldsymbol{u}=\begin{pmatrix}u_1\\u_2\\\vdots\\u_N\end{pmatrix},\quad \boldsymbol{v}=\begin{pmatrix}v_1\\v_2\\\vdots\\v_N\end{pmatrix},\quad \boldsymbol{\Phi}=\begin{pmatrix}\Phi_1\\\Phi_2\\\vdots\\\Phi_N\end{pmatrix}. \tag{13}$$

Obwohl kein Hinderungsgrund für die weitere Ableitung, nehmen wir der Einfachheit halber eine druckunabhängige statische Stabilität σ an, so daß die die vertikale Struktur des N-Flächen-Modells darstellende $N\times N$ Matrix $\boldsymbol{A}$ hier speziell

$$\boldsymbol{A}=\begin{pmatrix}1&-1&0&0\ldots&0&0\\-1&2&-1&0\ldots&0&0\\0&-1&2&-1\ldots&0&0\\\vdots&&&&&\\0&0&0&0\ldots&-1&1+\chi\end{pmatrix} \tag{14}$$

lautet. Durch die vertikale Diskretisierung tritt $\boldsymbol{A}$ an die Stelle von $-\frac{\partial^2}{\partial p^2}$. Von qualitativ wesentlicher Bedeutung ist hierbei die Zahl

$$\chi=\frac{\varrho\sigma\Delta p}{1+(\varrho\sigma\Delta p)/2}\approx 0{,}1\,,$$

die mathematisch gesehen die Singularität von $\boldsymbol{A}$ verhindert und physikalisch die durch die untere Randbedingung (6) bzw. (9) bedingte Existenz externer Gravitationswellen zum Ausdruck bringt. Wenn man externe Gravitationswellen ausfiltert, indem statt (6) $\omega=0$ gesetzt wird, so wird dieser Zustand formal auch durch $\chi=0$ erzielt.

Wir wenden nun die Transformation

$$\boldsymbol{T}\boldsymbol{A}\boldsymbol{T}^{-1}=\begin{pmatrix}\lambda_1&0&0&\ldots&0\\0&\lambda_2&0&\ldots&0\\\vdots&&\ddots&&\vdots\\0&0&0&\ldots&\lambda_N\end{pmatrix} \tag{15}$$

auf das Gleichungssystem (10)–(12) an, um einem wichtigen dynamischen Prinzip der baroklinen Ausgangsgleichungen zu entsprechen. Die Matrix $\boldsymbol{T}$ wird dabei so bestimmt, daß die $\lambda_1, \lambda_2, \ldots, \lambda_N$ ($\lambda_1<\lambda_2<\ldots<\lambda_N$) die Eigenwerte (sämtlich reell) von $\boldsymbol{A}$ sind. Die Vektoren $\boldsymbol{u}$, $\boldsymbol{v}$, $\boldsymbol{\Phi}$ transformieren sich in entsprechende Eigenvektoren

$$\boldsymbol{u}^T=\boldsymbol{T}\boldsymbol{u}=\begin{pmatrix}u_1^T\\u_2^T\\\vdots\\u_N^T\end{pmatrix},\quad \boldsymbol{v}^T=\boldsymbol{T}\boldsymbol{v}=\begin{pmatrix}v_1^T\\v_2^T\\\vdots\\v_N^T\end{pmatrix},\quad \boldsymbol{\Phi}^T=\boldsymbol{T}\boldsymbol{\Phi}=\begin{pmatrix}\Phi_1^T\\\Phi_2^T\\\vdots\\\Phi_N^T\end{pmatrix}, \tag{16}$$

bzw. in der Umkehrung gilt entsprechend

$$\boldsymbol{u}=\boldsymbol{T}^{-1}\boldsymbol{u}^T,\quad \boldsymbol{v}=\boldsymbol{T}^{-1}\boldsymbol{v}^T,\quad \boldsymbol{\Phi}=\boldsymbol{T}^{-1}\boldsymbol{\Phi}^T. \tag{17}$$

Das System (10)–(12) geht damit über in

$$\begin{gathered}\frac{\partial \boldsymbol{u}^T}{\partial t}+U\frac{\partial \boldsymbol{u}^T}{\partial x}-f\boldsymbol{v}^T+\frac{\partial \boldsymbol{\Phi}}{\partial x}=A\frac{\partial^2 \boldsymbol{u}}{\partial x^2}\\ \frac{\partial \boldsymbol{v}^T}{\partial t}+U\frac{\partial \boldsymbol{v}^T}{\partial x}+f\boldsymbol{u}^T=A\frac{\partial^2 \boldsymbol{v}^T}{\partial x^2}\\ \begin{pmatrix}\lambda_1&0&\ldots&0\\0&\lambda_2&&\\\vdots&&\ddots&\vdots\\0&&\ldots&\lambda_N\end{pmatrix}\left(\frac{\partial \boldsymbol{\Phi}^T}{\partial t}+U\frac{\partial \boldsymbol{\Phi}^T}{\partial x}-A\frac{\partial^2 \boldsymbol{\Phi}^T}{\partial x^2}\right)+\sigma(\Delta p)^2\frac{\partial \boldsymbol{u}^T}{\partial x}=0\end{gathered} \tag{18}$$

bzw. in Komponentenform, d. h. für die einzelnen Eigenfunktionen u_l^T, v_l^T, Φ_l^T ($l=1,2,\ldots,N$), geschrieben, in

$$\left.\begin{aligned}
&(19\text{a}) && \frac{\partial u_l^T}{\partial t} + U\frac{\partial u_l^T}{\partial x} - f v_l^T + \frac{\partial \Phi_l^T}{\partial x} = A\frac{\partial^2 u_l^T}{\partial x^2}\\
&(19\text{b}) && \frac{\partial v_l^T}{\partial t} + U\frac{\partial v_l^T}{\partial x} + f u_l^T = A\frac{\partial^2 v_l^T}{\partial x^2}\\
&(19\text{c}) && \lambda_l\left(\frac{\partial \Phi_l^T}{\partial t} + U\frac{\partial \Phi_l^T}{\partial x} - A\frac{\partial^2 \Phi_l^T}{\partial x^2}\right) + \sigma(\Delta p)^2\frac{\partial u_l^T}{\partial x} = 0
\end{aligned}\right\} l=1,\ldots,N$$

Wenn $\lambda_l \neq 0$, kann in der Prognosegleichung für die Eigenfunktionen des Geopotentials durch λ_l geteilt werden. Tatsächlich ist das bei berücksichtigten externen Gravitationswellen für alle λ_l ($l=1,\ldots,N$) der Fall, so daß

$$(19\text{c}') \qquad \frac{\partial \Phi_l^T}{\partial t} + U\frac{\partial \Phi_l^T}{\partial x} + c_l^2\frac{\partial u_l^T}{\partial x} = A\frac{\partial^2 \Phi_l^T}{\partial x^2}, \quad l=1,\ldots,N$$

folgt. $c_l = \pm\Delta p\sqrt{\frac{\sigma}{\lambda_l}}$ entspricht dabei den Phasengeschwindigkeiten der insgesamt N gravity modes des N-Flächen-Modells. Der kleinste Eigenwert λ_1 (bei der Anordnung ($\lambda_1 < \lambda_2 < \ldots < \lambda_N$)) entspricht demnach der schnellsten, externen Gravitationswelle. Man überzeugt sich durch einfaches Nachvollziehen der Rechnung, daß die untere Randbedingung $\omega=0$ anstelle von (6) externe Gravitationswellen ausfiltert. In der Matrix $\boldsymbol{A}$ ergibt sich $\chi=0$ mit $0<\lambda_2<$ $<\ldots<\lambda_N$, also $\lambda_1=0$. Für $l=1$ ist dann der Übergang von (19c) nach (19c′) nicht erlaubt. In (19) degeneriert (19c) zu der Filterbedingung $\frac{\partial u_1^T}{\partial x}=0$, was gemäß (16) das Verschwinden der über die Vertikale summierten Divergenzen bedeutet und gleichbedeutend mit $\omega(p=p_*)=0$ laut Kontinuitätsprinzip sein muß. Umgekehrt läßt sich nicht ohne Berechnung der Eigenwerte von $\boldsymbol{A}$ von $\omega(p=p_*)=0$ und über die vertikal integrierte Kontinuitätsgleichung mit Hilfe von (16) auf $\frac{\partial u_1^T}{\partial x}=0$ schließen, was nämlich mit (19c) $\lambda_1=0$ ergäbe! Aus (19) wird ersichtlich, daß die Prognose der Geopotentialeigenfunktion Φ_1^T wegen $\frac{\partial u_1^T}{\partial x}=0$ auf Umwegen über eine Randwertaufgabe durch Lösen einer Poissongleichung $\nabla^2\Phi_1^T=F$, die stellvertretend für $\frac{\partial u_1^T}{\partial x}=0$ als Filterbedingung fungiert, pro Zeitschritt gelöst werden muß. Die Lösung einer Poissongleichung impliziert aber eine unendlich große Fernwirkung, was das Verständnis von $c_1=\infty$ aus $\lambda_1=0$ fördert. Da Φ_1^T wegen (16) ebenfalls als vertikale Summe über die Geopotentialwerte aller N Flächen verstanden werden kann, ist mit der Lösung der genannten Poissongleichung der Zusammenhang mit der Integrationsmethode bei *Hinkelmann* (1959) offensichtlich.

Im folgenden verbleiben wir bei dem Fall $w_*=0$ (mit $\chi\neq 0$ und $\lambda_1\neq 0$). Nach der vertikalen Diskretisierung soll nun auch noch eine Diskretisierung zusätzlich in der Horizontalen (hier also längs der x-Achse) angenommen werden. Für den j-ten Gitterpunkt gilt dann $x_j=jd$, wenn d die Maschenweite darstellt. Für die Gitterpunktfunktion $\alpha_j=e^{ikjd}$ mit der Wellenzahl k gilt dabei mit $\nu=(\sin kd)/kd$

$$\frac{\alpha_{j+1}-\alpha_{j-1}}{2d} = ik\nu\alpha_j$$

$$\frac{\alpha_{j+2}+\alpha_{j-2}-2\alpha_j}{4d^2} = -k^2\nu^2\alpha_j\,.$$

Diese Diskretisierung in (19a, b, c′) eingeführt, liefert ein Wellenansatz mit der darauf nullgesetzten Koeffizientendeterminante die Phasengeschwindigkeiten

$$(20) \qquad \begin{aligned} \xi_1 &= -ik\nu^2A + \nu U\\ (\xi_{2,3})_l &= -ik\nu^2A + \nu U \pm \sqrt{\nu^2c_l^2 + f^2/k^2}\,, \end{aligned}$$

womit in bekannter Weise gezeigt ist, daß ein von Null verschiedener Diffusionskoeffizient eine Amplitudendämpfung bewirkt, und daß wegen $\nu<1$ bei $d>0$ die horizontale Diskretisierung grundsätzlich die tatsächlichen Signalgeschwindigkeiten in Abhängigkeit von der Wellenzahl verlangsamt darstellt. Wird schließlich das homogene Gleichungssystem gelöst, so läßt sich eine Normierung der Amplituden in geeigneter Weise vornehmen. Es ergibt sich schließlich als analytische Lösung für das horizontale diskretisierte System (19a, b, c′)

$$(21\text{a}) \qquad \begin{pmatrix} u_l^T\\ v_l^T\\ \Phi_l^T \end{pmatrix}_j = \begin{pmatrix} 0\\ \frac{ik\nu}{f}\\ 1 \end{pmatrix}(\Phi_{l,1}^T)_j + \begin{pmatrix} \frac{\sqrt{\nu^2c_l^2+f^2/k^2}}{\nu c_l^2}\\ -\frac{if/k}{\nu c_l^2}\\ 1 \end{pmatrix}(\Phi_{l,2}^T)_j + \begin{pmatrix} -\frac{\sqrt{\nu^2c_l^2+f^2/k^2}}{\nu c_l^2}\\ -\frac{if/k}{\nu c_l^2}\\ 1 \end{pmatrix}(\Phi_{l,3}^T)_j \quad (\text{für } l=1,\ldots,N),$$

wobei die Eigenfunktionen allein auf die entsprechenden Komponenten der Eigenfunktionen des Geopotential zurückgeführt sind. Diese lauten im Falle der hier belassenen zeitlichen Ableitungen in differentieller Form

$$(21\text{b}) \qquad \left.\begin{aligned} (\Phi_{l,1}^T)_j &= \overset{*}{\Phi}{}_{l,1}^T\, e^{-k^2\nu^2At}\, e^{ik(jd-\nu Ut)}\\ (\Phi_{l,\frac{2}{3}}^T)_j &= \overset{*}{\Phi}{}_{l,\frac{2}{3}}^T\, e^{-k^2\nu^2At}\, e^{ik[jd-(\nu U\pm\sqrt{\nu^2c_l^2+f^2/k^2})t]} \end{aligned}\right\} l=1,\ldots,N$$

Mit (21) ist wegen (17) schließlich auch die analytische Lösung des vertikal und horizontal diskretisierenden Ausgangsgleichungssystems (1)–(6) gegeben. Die $\overset{*}{\Phi}{}^T_{l,1}$, $\overset{*}{\Phi}{}^T_{l,2}$, $\overset{*}{\Phi}{}^T_{l,3}$ sind nur insofern frei wählbar, als sie durch eine sinngemäße Initialisierung festgelegt werden müssen. Die linearen Ausgangsgleichungen lassen sich demzufolge zurückführen auf folgende Gleichungen

$$\frac{\partial(\Phi^T_{l,1})_j}{\partial t}+ik\nu U(\Phi^T_{l,1})_j=-k^2\nu^2 A(\Phi^T_{l,1})_j \tag{22}$$

$$\frac{\partial(\Phi^T_{l,\frac{2}{3}})_j}{\partial t}+ik\nu U(\Phi^T_{l,\frac{2}{3}})_j\pm ik\sqrt{\nu^2c_l^2+f^2/k^2}\,(\Phi^T_{l,\frac{2}{3}})_j=-k^2\nu^2A(\Phi^T_{l,\frac{2}{3}})_j$$

$$(l=1,\ldots,N)$$

in Einklang mit (21) und jeder der drei Gleichungen (19a, b, c'). Wir dürfen dieses Ergebnis in Anlehnung an *Kurihara* (1965) als „Kuriharagleichungen“ bezeichnen, wobei der genannte Autor anhand von solchen typischen Gleichungen wie (22) verschiedene Zeitdifferenzenschemata untersuchte. Mit dem wichtigen Resultat (22) sind wir folglich gleichermaßen in der Lage, für das barokline N-Flächen-Modell verschiedene Zeitschemata linear zu analysieren.

3. Ein semi-implizites Zeitschema

Hier ist beabsichtigt, an dem in Abschn. 2 beschriebenen linearen Modell als ein Anwendungsbeispiel ein semi-implizites Zeitschema zu untersuchen. Das Zeitdifferenzenschema soll nur zwei Zeitniveaus benutzen. Für die Ausgangsgleichungen (1)–(4) ergibt sich damit

$$u^{(\tau+1)}=u^{(\tau)}-\Delta t\left(U\frac{\partial u^{(\tau)}}{\partial x}-fv^{(\tau+1)}+\frac{\partial\Phi^{(\tau+1)}}{\partial x}-A\frac{\partial^2u^{(\tau)}}{\partial x^2}\right) \tag{23}$$

$$v^{(\tau+1)}=v^{(\tau)}-\Delta t\left(U\frac{\partial v^{(\tau)}}{\partial x}+fu^{(\tau+1)}-A\frac{\partial^2 v^{(\tau)}}{\partial x^2}\right) \tag{24}$$

$$\Psi^{(\tau+1)}=\Psi^{(\tau)}-\Delta t\left(U\frac{\partial\Psi^{(\tau)}}{\partial x}+\frac{\partial u^{(\tau+1)}}{\partial x}-A\frac{\partial^2\Psi^{(\tau)}}{\partial x^2}\right) \tag{25}$$

mit $\Psi=-\frac{\partial}{\partial p}\left(\frac{1}{\sigma}\frac{\partial\Phi}{\partial p}\right)$. Man erkennt aus den Rechnungen in Abschn. 2, daß das formulierte Zeitschema schließlich in folgender Weise für die Kuriharagleichungen (22) angewendet werden muß, wenn externe Gravitationswellen enthalten sind und die vertikale und horizontale Diskretisierung gleichfalls angenommen wird:

$$(\Phi^T_{l,1})^{(\tau+1)}_j=(\Phi^T_{l,1})^{(\tau)}_j-i\beta(\Phi^T_{l,1})^{(\tau)}_j-\gamma(\Phi^T_{l,1})^{(\tau)}_j \tag{26}$$

$$(\Phi^T_{l,\frac{2}{3}})^{(\tau+1)}_j=(\Phi^T_{l,\frac{2}{3}})^{(\tau)}_j\pm i\alpha(\Phi^T_{l,\frac{2}{3}})^{(\tau+1)}_j-i\beta(\Phi^T_{l,\frac{2}{3}})^{(\tau)}_j-\gamma(\Phi^T_{l,\frac{2}{3}})^{(\tau)} \tag{27}$$

$$(l=1,\ldots,\mathrm{N})$$

Dabei bedeuten

$$\alpha=k\Delta t\sqrt{\nu^2c_l^2+f^2/k^2}$$

$$\beta=k\Delta t\nu U$$

$$\gamma=k^2\nu^2A\Delta t\,.$$

Nach dem Muster der Überlegungen in Abschn. 2 läßt sich hier, von (23)–(25) ausgehend, ganz analog vorgehen bis zu einschließlich der Beziehung (21a) für die zeitlich diskretisierten Eigenfunktionen mit deren zweckmäßiger Normierung bzw. Zurückführung auf allein die Rossby- bzw. Gravitationswellen-Anteile $(\Phi^T_{l,1})^{(\tau)}_j$ bzw. $(\Phi^T_{l,2})^{(\tau)}_j$, $(\Phi^T_{l,2})^{(\tau)}_j$ der Geopotentialeigenfunktionen. Insofern läßt sich auch die analytische Lösung des vertikal und horizontal diskretisierten Modells (23)–(25) für das semi-implizite Schema mit Hilfe von (17) angeben, wenn nur die diesem Zeitschema entsprechend abgeänderten Relationen (21b) aufgestellt sind. Wenn man die Rechnungen in Abschn. 2 verfolgt, so wird für den Übergang von (23)–(25) zu den entsprechenden Kuriharagleichungen (26)–(27) deutlich, daß es für die Rechnungen bequem und klar erscheint, wenn sowohl der Druckgradientterm als auch der Coriolisterm zum gleichen Zeitniveau im Gegensatz zu vielen semi-impliziten Formulierungen angeschrieben sind. Dies wird physikalisch deshalb verständlich, weil diese beiden Terme vom Gesichtspunkt der Adaption her als unbedingt zusammengehörig betrachtet werden müssen.

Um endlich Beziehungen in Analogie zu (21b) aufstellen zu können, sind die dem speziellen Zeitschema eigenen Amplituden und Phasen zu bestimmen. Die Gleichungen (26)–(27) schreiben sich auch als

$$(\Phi^T_{l,1})^{(\tau+1)}_j=(1-\gamma-i\beta)(\Phi^T_{l,1})^{(\tau)}_j=|A_1|e^{i\Psi_1}(\Phi^T_{l,1})^{(\tau)}_j$$

$$(\Phi^T_{l,\frac{2}{3}})^{(\tau+1)}_j=\left(\frac{1-\gamma-i\beta}{1\pm i\alpha}\right)(\Phi^T_{l,\frac{2}{3}})^{(\tau)}_j=|A_{l,\frac{2}{3}}|e^{i\Psi_{l,\frac{2}{3}}}(\Phi^T_{l,\frac{2}{3}})^{(\tau)}_j$$

$$(l=1,\ldots,N)$$

mit den Amplituden- und Phasenänderungen pro Zeitschritt für den für jedes l gleichen Rossby-Teil

$$|\Lambda_1| = \sqrt{(1-\gamma)^2+\beta^2}\,, \quad \Psi_1 = -\arctan\left(\frac{\beta}{1-\gamma}\right)$$

und für den wegen $\alpha = \alpha_l (c_1 > c_2 > \ldots > c_N)$ von l abhängigen Trägheits- und Gravitations-Teil

$$|\Lambda_{l,2}| = |\Lambda_{l,3}| = \sqrt{\frac{(1-\gamma)^2+\beta^2}{1+\alpha^2}}\,, \quad \Psi_{l,{2 \atop 3}} = -\arctan\left(\frac{\beta \pm \alpha(1-\gamma)}{(1-\gamma) \mp \alpha\beta}\right).$$

Die von (21 b) analogen Beziehungen lauten damit

$$(\Phi^T_{l,1})^{(\tau)}_j = \left(\sqrt{(1-k^2\nu^2 A\Delta t)^2 + k^2\nu^2\Delta t^2 U^2}\right)^\tau \overset{*}{\Phi}{}^T_{l,1}\, e^{ik\left[jd - \frac{\tau}{k}\arctan\left(\frac{k\nu\Delta t U}{1-k^2\nu^2\Delta t A}\right)\right]} \tag{29}$$

$$(\Phi^T_{l,{2\atop 3}})^{(\tau)}_j = \left(\sqrt{\frac{(1-k^2\nu^2 A\Delta t)^2 + k^2\nu^2\Delta t^2 U^2}{1+k^2\Delta t^2(\nu^2 c_l^2 + f^2/k^2)}}\right)^\tau \overset{*}{\Phi}{}^T_{l,{2\atop 3}}\, e^{ik\left[jd - \frac{\tau}{k}\arctan\left(\frac{k\Delta t(\nu U \pm \sqrt{\nu^2 c_l^2 + f^2/k^2}(1-k^2\nu^2\Delta t A))}{(1-k^2\nu^2\Delta t A) \mp k^2\nu^2\Delta t^2 U\sqrt{\nu^2 c_l^2+f^2/k^2}}\right)\right]} \tag{30}$$

Um die Bedeutung des eingeführten lateralen Diffusionskoeffizienten A zu verdeutlichen, sei dieser zunächst gleich Null angenommen. Für die Rossby-Physik bedeutet aber dann

$$\left(\sqrt{1+k^2\nu^2\Delta t^2 U^2}\right)^\tau > 1$$

prinzipiell eine zeitlich zunehmende Amplitudenvergrößerung, also numerisch bedingte, wenn auch schwache Instabilität, die für kürzere Wellenlängen mehr ins Gewicht fällt als für längere. Für Trägheits-Gravitationswellen hingegen ergibt sich

$$\left(\sqrt{\frac{1+k^2\nu^2\Delta t^2 U^2}{1+k^2\Delta t^2(\nu^2 c_l^2+f^2/k^2)}}\right)^\tau < 1\,,$$

wenn nur $U^2 < \nu^2 c_l^2 + f^2/k^2$ vorausgesetzt wird, d. h. diese Wellenklasse wird gedämpft. Der Grad der Dämpfung ist dabei offensichtlich für die schnellsten Gravitationswellen am wirksamsten. Es steht die Frage, ob für die langsamsten internen Gravitationswellen in einem N-Flächen-Modell mit sehr hoher vertikaler Auflösung bei der Simulation mesoskaler Effekte, für die der Trägheitseffekt vernachlässigbar wird, Instabilität möglich sein kann, falls nämlich mindestens die Situation $\nu^2 U^2 > \nu^2 c_N^2 + f^2/k^2$ eintritt. Im allgemeinen ist aber dieser Fall auszuschließen.

Insbesondere aus der Absicht, die Instabilität im Rossbyregime nicht zuzulassen, erwächst die Bedeutung eines von Null verschiedenen Diffussionskoeffizienten. A ist offensichtlich so zu bestimmen, daß $|\Lambda_1| = 1$ bewirkt wird. Das wird erreicht mit

$$k^2\nu^2\Delta t A = 1 \overline{(+)} \sqrt{1-k^2\nu^2\Delta t^2 U^2}\,. \tag{31}$$

A ist insofern nicht doppeldeutig, als der positive Wert der Wurzel ausscheiden muß, um nicht vorzeichenverkehrte, falsche Phasengeschwindigkeiten zuzulassen. Weiterhin wird aus (31) das verbindliche Stabilitätskriterium $k^2\nu^2\Delta t^2 U^2 \leqq 1$ bzw.

$$|U|\frac{\Delta t}{d} \leqq 1 \tag{31'}$$

erkenntlich.

4. Eine quantitative Analyse für den Spezialfall $N=2$ mit $\chi=0$

Um quantitative Vorstellungen von der stabilisierenden Wirkung von A bezüglich des Rossby-Regimes im Zusammenhang mit dem zur Diskussion stehenden semi-impliziten Zeitschema zu erhalten, ist im folgenden als Spezialfall ein 2-Flächenmodell mit ausgefilterten externen Gravitationswellen ausreichend. Seine Vertikalstruktur ist in Abb. 2 wiedergegeben.

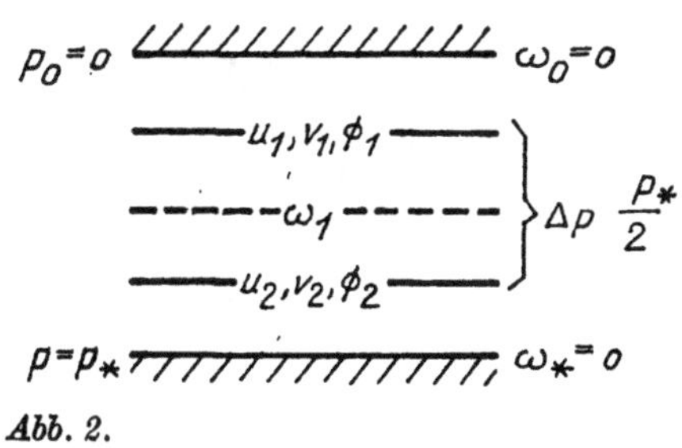

Abb. 2.

Die Matrix $\boldsymbol{A}$ lautet mit $\chi=0$

$$\boldsymbol{A} = \begin{pmatrix} 1 & -1 \\ -1 & 1 \end{pmatrix},$$

ihre Eigenwerte sind $\lambda_1=0$, $\lambda_2=2$. Die Matrizen $\boldsymbol{T}$, $\boldsymbol{T}^{-1}$ nehmen die spezielle Gestalt

$$\boldsymbol{T}=\begin{pmatrix}1 & 1\\ 1 & -1\end{pmatrix}, \quad \boldsymbol{T}^{-1}=\begin{pmatrix}\frac{1}{2} & \frac{1}{2}\\ \frac{1}{2} & -\frac{1}{2}\end{pmatrix}$$

an, und (17) lautet speziell

$$u_1=\frac{1}{2}(u_1^T+u_2^T), \quad v_1=\frac{1}{2}(v_1^T+v_2^T), \quad \Phi_1=\frac{1}{2}(\Phi_1^T+\Phi_2^T) \tag{32}$$

$$u_2=\frac{1}{2}(u_1^T-u_2^T), \quad v_2=\frac{1}{2}(v_1^T-v_2^T), \quad \Phi_2=\frac{1}{2}(\Phi_1^T-\Phi_2^T).$$

Anstelle von (21a) finden wir für den mode $l=1$ aufgrund der durch $\chi=0$ ausgefilterten externen Gravitationswellen

$$\begin{pmatrix}u_1^T\\ v_1^T\\ \Phi_1^T\end{pmatrix}_j=\begin{pmatrix}0\\ \frac{ik\nu}{f}\\ 1\end{pmatrix}(\Phi_{1,1}^T)_j.$$

Das ist unmittelbar eine Folge von $\frac{\partial u_1^T}{\partial x}=0$ in dem System (19a–c) für $l=1$ wegen $\lambda_1=0$. Für den mode $1=2$ sind die Gravitationswellenkomponenten vorhanden:

$$\begin{pmatrix}u_2^T\\ v_2^T\\ \Phi_2^T\end{pmatrix}=\begin{pmatrix}0\\ \frac{ik\nu}{f}\\ 1\end{pmatrix}(\Phi_{2,1}^T)_j+\begin{pmatrix}\frac{\sqrt{\nu^2c_2^2+f^2/k^2}}{\nu c_2^2}\\ -\frac{if/k}{\nu c_2^2}\\ 1\end{pmatrix}(\Phi_{2,2}^T)_j+\begin{pmatrix}-\frac{\sqrt{\nu^2c_2^2+f^2/k^2}}{\nu c_2^2}\\ -\frac{if/k}{\nu c_2^2}\\ 1\end{pmatrix}(\Phi_{2,3}^T)_j.$$

Während $c_1=\infty$, gilt $c_2=58\ \mathrm{ms}^{-1}$ (bei entsprechendem numerischen Wert von σ). Mit (32) lautet bei bekannten Komponenten der Geopotentialeigenfunktionen dann die Lösung für die ursprünglichen Funktionen

$$\begin{aligned}
u_{1,j}^{(\tau)}&=\frac{\sqrt{\nu^2c_2^2+f^2/k^2}}{2\nu c_2^2}(\Phi_{2,2}^T-\Phi_{2,3}^T)_j^{(\tau)}\\
u_{2,j}^{(\tau)}&=-\frac{\sqrt{\nu^2c_2^2+f^2/k^2}}{2\nu c_2^2}(\Phi_{2,2}^T-\Phi_{2,3}^T)_j^{(\tau)}\\
v_{1,j}^{(\tau)}&=\frac{ik\nu}{2f}(\Phi_{1,1}^T+\Phi_{2,1}^T)_j^{(\tau)}-\frac{if/k}{2\nu c_2^2}(\Phi_{2,2}^T+\Phi_{2,3}^T)_j^{(\tau)}\\
v_{2,j}^{(\tau)}&=\frac{ik\nu}{2f}(\Phi_{1,1}^T-\Phi_{2,1}^T)_j^{(\tau)}+\frac{if/k}{2\nu c_2^2}(\Phi_{2,2}^T+\Phi_{2,3}^T)_j^{(\tau)}\\
\Phi_{1,j}^{(\tau)}&=\frac{1}{2}(\Phi_{1,1}^T+\Phi_{2,1}^T)_j^{(\tau)}+\frac{1}{2}(\Phi_{2,2}^T+\Phi_{2,3}^T)_j^{(\tau)}\\
\Phi_{2,j}^{(\tau)}&=\frac{1}{2}(\Phi_{1,1}^T-\Phi_{2,1}^T)_j^{(\tau)}-\frac{1}{2}(\Phi_{2,2}^T+\Phi_{2,3}^T)_j^{(\tau)}\\
\omega_{1,j}^{(\tau)}&=-ik\Delta p\,\frac{\sqrt{\nu^2c_2^2+f^2/k^2}}{2c_2^2}(\Phi_{2,2}^T-\Phi_{2,3}^T)_j^{(\tau)}.
\end{aligned} \tag{33}$$

Aus den für $N=2$ mit $\chi=0$ spezialisierten Kuriharagleichungen ergibt sich nun bei Anwendung des semi-impliziten Zeitschemas für die einzelnen Komponenten der Geopotentialeigenfunktionen für den Rossby-Anteil

$$\begin{pmatrix}\Phi_{1,1}^T\\ \Phi_{2,1}^T\end{pmatrix}_j^{(\tau)}=\left(\sqrt{(1-k^2\nu^2A\Delta t)^2+k^2\nu^2\Delta t^2U^2}\right)^\tau\begin{pmatrix}\mathring{\Phi}_{1,1}^T\\ \mathring{\Phi}_{2,1}^T\end{pmatrix}e^{ik\left[jd-\frac{\tau}{k}\arctan\left(\frac{k\nu\Delta tU}{1-k^2\nu^2A\Delta t}\right)\right]} \tag{34}$$

und für den Trägheits-Gravitationswellen-Anteil

$$\begin{pmatrix}\Phi_{2,2}^T\\ \Phi_{2,3}^T\end{pmatrix}_j^{(\tau)}=\left(\sqrt{\frac{(1-k^2\nu^2A\Delta t)^2+k^2\nu^2\Delta t^2U^2}{1+k^2\Delta t^2(\nu^2c_2^2+f^2/k^2)}}\right)^\tau\begin{pmatrix}\delta_1 & 0\\ 0 & \delta_2\end{pmatrix}\begin{pmatrix}\mathring{\Phi}_{2,2}^T\\ \mathring{\Phi}_{2,3}^T\end{pmatrix}. \tag{35}$$

$$\delta_{1,2}=e^{ik\left[jd-\frac{\tau}{k}\arctan\left(\frac{k\Delta t(\nu U\pm\sqrt{\nu^2c_l^2+f^2/k^2}(1-k^2\nu^2\Delta tA))}{(1-k^2\nu^2\Delta tA)\mp k^2\nu^2\Delta t^2U\sqrt{\nu^2c_2^2+f^2/k^2}}\right)\right]}$$

(34) und (35) geben im Zusammenhang mit (33) die numerischen Auswirkungen des semi-impliziten Schemas auf das 2-Flächenmodell wieder.
Im folgenden bestimmen wir das Amplitudenverhalten für das Rossby-Regime,

$$|\Lambda_1|^\tau = \left(\sqrt{(1-k^2\nu^2 A\Delta t)^2 + k^2\nu^2\Delta t^2 U^2}\right)^\tau, \tag{36}$$

für den Fall $A=0$.

Tabelle 1

n	$\lvert\Lambda_1\rvert^{24}$	$\lvert\Lambda_1\rvert^{48}$
2	1.	1.
4	2.070	4.284
6	1.733	3.002
10	1.292	1.670
14	1.151	1.324
20	1.074	1.153

In Tab. 1 sind für $|U|\Delta t/d = 0.25$, d. h. beispielsweise für $U = 75$ km h^{-1} mit $d = 300$ km, $\Delta t = 1$ h oder $d = 150$ km, $\Delta t = 0.5$ h etc., und $A = 0$ die Stabilitätsverhältnisse für den Rossby-Bereich in Abhängigkeit von der Wellenlänge in Gittereinheiten dargestellt. Es ist das Instabilitätsverhalten für kürzere Wellen, insbesondere für $4d$-Wellen, bemerkenswert, was zur Abschwächung den Übergang zu einem horizontal feinerem Gitter geradezu gebietet. Für längere Wellen ist ein Amplitudenwachstum prinzipiell vorhanden, so daß nur kurzfristige Prognosen sinnvoll sein können.

Um diesen Defekt zu beseitigen, wäre in (36) ein von Null verschiedener Diffusionskoeffizient gemäß (31) einzuführen. In Tab. 2 ist die Berechnung von A bzw. $A\Delta t/d^2$ für $|U|\Delta t/d = 0.25$ mit $U = 75$ km h^{-1} in Abhängigkeit von n wiedergegeben:

Tabelle 2

n	$A\Delta t/d^2$	A (für $d=300$ km, $\Delta t=1$ h)	$A\left(\text{für } d=150 \text{ km}, \Delta t=\frac{1}{2}\text{ h}\right)$
2	$3.125\cdot 10^{-2}$	$2.8125\cdot 10^3$ km^2 h^{-1}	$1.4063\cdot 10^3$ km^2 h^{-1}
4	3.175	2.8579	1.4289
6	3.154	2.8385	1.4193
10	3.140	2.8262	1.4131
14	3.134	2.8203	1.4101
20	3.130	2.8166	1.4083

Hierbei ist zu beachten, daß nach zweimaliger Anwendung der l'Hospitalschen Regel für $2d$-Wellen und in gleicher Weise für unendlich lange Wellen

$$\frac{\Delta t A_0}{d^2} = \frac{1}{2}\left(U\frac{\Delta t}{d}\right)^2 \quad \text{bzw.} \quad A_0 = \frac{U^2}{2}\Delta t \tag{37}$$

folgt. Mit diesem wellenlängenabhängigen A ist der wünschenswerte Zustand $|\Lambda_1|^\tau = 1$ streng erreicht. In praxi ist aber die Wellenlängenabhängigkeit von A unbequem, und man muß sich auf einen konstanten Wert bzw. auf den Wert von A für eine konstante Wellenlänge festlegen. Entscheidet man sich willkürlich für den in (37) angegebenen Grenzwert, so folgt aus (36)

$$|\Lambda_1^{(0)}|^\tau = \left(\sqrt{1+\frac{1}{4}\left(U\frac{\Delta t}{d}\right)^4 \sin^4(kd)}\right)^\tau = 1 + 0\,(10^{-2}+10^{-4})$$

d. h., über das ganze Spektrum ist mit hoher Genauigkeit $|\Lambda_1^{(0)}|^\tau = 1$ abgesichert.

Das anschließend mitgeteilte Ergebnis des Dämpfungsverhaltens für den Trägheits-Gravitationswellen-Anteil bezieht sich in Hinblick auf die erste Version eines Regionalmodells auf ein 4-Schichten-Modell ($N=4$). Die entsprechende Auswertung einer die Vertikalstruktur beschreibenden Matrix, die im Gegensatz zu (14) eine ganz bestimmte vertikal abhängige statische Stabilität berücksichtigt, liefert mit Einbeziehung externer Wellen die Phasengeschwindigkeiten $c_1 = \pm 256$ ms^{-1}, $c_2 = \pm 41$ ms^{-1}, $c_3 = \pm 22$ ms^{-1}, $c_4 = \pm 14$ ms^{-1}. Der relativ niedrige Wert für c_1 erklärt sich aus der Wahl der Modellobergrenze bei 200 mbar. In Tab. 3 ist das Ergebnis der Berechnung der Dämpfungsrate unter Verwendung der Wirkung von A_0

$$|\Lambda_{2,3}^{(0)}| = \frac{1}{\sqrt{1+\sin^2(kd)\left(\frac{\Delta t c_K}{d}\right)^2 + f^2\Delta t^2}}, \quad k=1,\ldots,4 \tag{38}$$

für $f = 0.5$ h^{-1} wiedergegeben. In der dargestellten Abhängigkeit von n und c_k bedeutet jeweils der obere Wert das Resultat für $d = 300$ km, $\Delta t = 1$ h und der untere Wert das Ergebnis für $d = 150$ km, $\Delta t = 0.5$ h. Für jedes k ergibt sich die stärkste Dämpfung für $4d$-Wellen, die geringste für $2d$-Wellen und sehr lange Wellen ($n\to\infty$). Grundsätzlich ist die Dämpfung für schnellere Wellen größer. Für eine feinere Gitterauflösung mit dem kleineren Zeitschritt ist die Dämpfung geringer. Die Dämpfung für $2d$-Wellen und unendlich lange Wellen ist gleich und nur dann kleiner Eins, wenn der Trägheitseffekt wesentlich wird. Die Dämpfung nähert sich diesem „Sättigungswert"

$$|\Lambda_{2,3}^{*}| = \frac{1}{\sqrt{1+f^2\Delta t^2}} \tag{39}$$

Tabelle 3

n	c_1	c_2	c_3	c_4
2	0.8944 0.9701	0.8944 0.9701	0,8944 0.9701	0,8944 0.9701
4	0,3059 0.3086	0,8187 0.8755	0.8705 0.9398	0.8845 0.9575
6	0.3465 0.3505	0.8357 0.8966	0.8763 0.9471	0.8869 0.9606
10	0.4709 0.4810	0.8660 0.9341	0.8859 0.9593	0.8910 0.9657
20	0.6818 0.7136	0.8863 0.9598	0.8921 0.9671	0.8935 0.9689
∞	0.8944 0.9701	0.8944 0.9701	0.8944 0.9701	0.8944 0.9701

im gesamten Spektralbereich bei hoher Vertikalauflösung, d. h. bei großem N mit sehr kleinem c_N. Mit $f=0.5\ \mathrm{h}^{-1}$ und $\Delta t=0.25$ h ($\triangleq d=75$ km) gilt beispielsweise $|\Lambda^*_{2.3}|=0.9923$. Daß $4d$-Wellen, die wie $2d$-Wellen zum Rechenlärm-Bereich zählen, am stärksten gedämpft werden, und, daß dieses besonders für externe Wellen gilt, ist erwünscht. Da aber $2d$-Wellen am wenigsten unterdrückt werden, ist für diese u. U. eine Sonderbehandlung erforderlich. In der trockenen Regionalvariante wie auch im linearen Experiment hat sich aber bestätigt, daß $2d$-Wellen allein von der seitlichen Berandung herrühren, und, dort bekämpft, im gesamten Feld verschwinden, so daß die relative Durchlaßfähigkeit des Zeitschemas für $2d$-Wellen keine Konsequenzen hat.

5. Schlußbemerkungen

In einer linearen Analyse untersuchen wir ein aus ökonomischen Gründen wünschenswertes semi-implizites Zeitschema für ein baroklines Modell. Wir stellen die folgenden Eigenschaften desselben heraus. 1. Das semi-implizite Schema verwendet nur zwei Zeitniveaus und ist deshalb aus speicherökonomischen Gründen von Vorteil. 2. Das Zeitschema läßt einen rein numerisch bedingten, fiktiven Lösungsanteil nicht zu. 3. Das Schema ist selektiv dämpfend insofern, als nur der Lösungsanteil bezüglich der Trägheits- und Gravitationswellen gedämpft wird. Die größte Dämpfung erfahren dabei die schnellsten Wellen. 4. Der Rossby-Bereich der Lösung wird durch die geeignete Wahl eines lateralen Diffusionskoeffizienten praktisch vollkommen erhalten. 5. Die durch das Zeitschema bedingten Phasenfehler sind gering. Letzteres begründet sich auf einen hier im einzelnen nicht mitgeteilten quantitativen Vergleich der bekannten ν-Kurve, die den relativen Fehler der Phasengeschwindigkeit U in Abhängigkeit von n allein durch die räumliche Gitterwirkung zum Ausdruck bringt $\left(\nu=\frac{\nu U}{U}\right)$, gegenüber dem relativen Fehler bei räumlicher und zeitlicher Diskretisierung

$$\varepsilon_1=\frac{\frac{1}{k\Delta t}\arc\tan\left(\frac{k\nu\Delta tU}{1-k^2\nu^2\Delta tA_0}\right)}{U}$$

und einem Vergleich der Kurven

$$\frac{\nu U\pm\sqrt{\nu^2c_2^2+f^2/k^2}}{U\pm\sqrt{c_2^2+f^2/k^2}}$$

mit

$$\varepsilon_{2,3}=\frac{\frac{1}{k\Delta t}\arc\tan\left(\frac{k\Delta t(\nu U\pm\sqrt{\nu^2c_2^2+f^2/k^2}\,(1-k^2\nu^2A_0\Delta t))}{(1-k^2\nu^2A_0\Delta t)\mp k^2\Delta t^2\nu U\sqrt{\nu^2c_2^2+f^2/k^2}}\right)}{U\pm\sqrt{c_2^2+f^2/k^2}}.$$

Abschließend sei mitgeteilt, daß auf der Grundlage dieser Überlegungen ein semi-implizites Regionalmodell realisiert wurde, welches in seinem vorläufigen Zustand 4-flächig rechnet, ein Gebiet von 23×27 Gitterpunkten überdeckt und bei einer Gitterweite von z. Z. $d=300$ km einen Zeitschritt von 1 h verwendet. Die seitliche Berandung ist offen, und Resultate einer Hemisphärenprognose liefern ohne feed back die nötigen Randbedingungen. Darüber wird in einer folgenden Mitteilung berichtet.

Literatur

[1] *Bortnikov, S. A.*: Opyt kratkosročnogo prognoza pogody na osnove rešenija polnoj systemy uravnenij termogidrodinamiki. Meteorol. i Gidrol. 1962, № 11.
[2] *Hinkelmann, K.*: Ein numerisches Experiment mit den primitiven Gleichungen. The Rossby Memorial Volume, Rockefeller Institute Press 1959.
[3] *Kurihara, Y* : On the use of implicit and iterative methods for the time integration of the wave equation. Month. Weath. Rev. 93 (1965) 1.
[4] *Kwizak, M.* und *A. J. Robért*: A semi-implicit scheme for grid point atmospheric modells of the primitive equations. Month. Weath. Rev. 99 (1971) 1.
[5] *Marčuk, G. I.*: Čislennye metody v prognoze pogody. Gidrometeoizdat, Leningrad 1967.
[6] *Miyakoda, K.*: Contribution to the numerical weather prediction-computation with finite difference. Japanese Journal of Geophysics 3 (1962) 1.
[7] *Robért, A. J.*: The integration of a spectral model of the atmosphere by the implicit method. Proc. of the WMO/IUGG-Symposium on numerical weather prediction, Tokyo, Nov. 1968.
[8] *Thompson, P. D.*: Numerical Weather Analysis and Prediction. New York 1961.

Anschrift: Dr. H.-J. Herzog, Meteorologischer Dienst der DDR, Zentrale Wetterdienststelle, DDR–15 Potsdam, Michendorfer Chaussee 23

551.509.313

Meteorologischer Sonderbericht

Ergänzende Bemerkung zur Wahl seitlicher Randbedingungen für ein barotropes Modell anhand eines linearisierten Gleichungssystems

Von A. Meyer

Mit 1 Abbildung

1. Einleitung

Es ist allgemein bekannt, daß dem Problem der seitlichen Randbedingungen besondere Aufmerksamkeit gewidmet werden muß, insbesondere dann, wenn die Praxis ein seitlich begrenztes Rechengebiet fordert. Dies trifft in besonderem Maße für nationale meteorologische Zentren (NMC) zu. Obgleich bezüglich dieser Problemstellung speziell in letzterer Zeit Aktivitäten festzustellen waren und noch sind, läßt sich ein gewisser theoretischer Nachholebedarf nicht übersehen. Ungeachtet dieser Tatsache findet man einige pragmatische Lösungen, die sich als recht brauchbar erwiesen haben. Demgegenüber steht die Schwierigkeit, gefundene theoretische Ergebnisse in die Praxis zu überführen. Trotz alledem scheint es gewiß zweckmäßig, in theoretischen Untersuchungen fortzufahren. In diesem Sinne soll dieser Artikel als eine Ergänzung zu der Arbeit von *Herzog* und *Meyer* (1976) verstanden werden.

2. Ableitung von Randintegralbeziehungen

Ausgangspunkt der folgenden Untersuchungen sei ein linearisiertes Gleichungssystem, in dem die Variablen u, v und h Störgrößen darstellen mögen.

$$\begin{aligned} &\frac{\partial u}{\partial t}+U\frac{\partial u}{\partial x}+V\frac{\partial u}{\partial y}+g\frac{\partial h}{\partial x}=0\\ &\frac{\partial v}{\partial t}+U\frac{\partial v}{\partial x}+V\frac{\partial v}{\partial y}+g\frac{\partial h}{\partial y}=0\\ &\frac{\partial h}{\partial t}+U\frac{\partial h}{\partial x}+V\frac{\partial h}{\partial y}+H\left\{\frac{\partial u}{\partial x}+\frac{\partial v}{\partial y}\right\}=0 \end{aligned} \tag{1}$$

U und V seien die Komponenten einer konstanten Grundströmung und H eine konstante mittlere Höhe.

Es erweist sich als recht zweckmäßig für weitere Überlegungen das Gleichungssystem (1) in eine entsprechende Form zu transformieren.

$$\begin{aligned} &\frac{\partial \Phi_1}{\partial t}+U\frac{\partial \Phi_1}{\partial x}+V\frac{\partial \Phi_1}{\partial y}=0\\ &\frac{\partial \Phi_2}{\partial t}+U\frac{\partial \Phi_2}{\partial x}+V\frac{\partial \Phi_2}{\partial y}+\sqrt{gH}\,\nabla\vec{\Phi}_3=0\\ &\frac{\partial \vec{\Phi}_3}{\partial t}+U\frac{\partial \vec{\Phi}_3}{\partial x}+V\frac{\partial \vec{\Phi}_3}{\partial y}+\sqrt{gH}\,\nabla\Phi_2=0 \end{aligned} \tag{2}$$

wobei die Variablen Φ_1, Φ_2 und $\vec{\Phi}_3$ folgender Gestalt sind

$$\Phi_1=\frac{\partial v}{\partial x}-\frac{\partial u}{\partial y},\quad \Phi_2=H\left\{\frac{\partial u}{\partial x}+\frac{\partial v}{\partial y}\right\},\quad \vec{\Phi}_3=\sqrt{gH}\,\nabla h. \tag{3}$$

Die erste Gleichung des Gleichungssystems (2), gewissermaßen die Vorticitygleichung, stellt den Rossbyanteil des Systems dar, währenddessen die beiden anderen miteinander verkoppelten Gleichungen den Gravitationsanteil ausmachen. Für die auf diese Weise entkoppelte erste Gleichung läßt sich relativ einfach eine Randintegralbeziehung unter Anwendung der Energie-Integral-Methode finden. Aus

$$\frac{\partial \Phi_1}{\partial t}+\mathfrak{v}\,\nabla\Phi_1=0;\quad \mathfrak{v}=u_i+v_j$$

folgt mit Hilfe des Gaußschen Integralsatzes

$$\frac{\partial}{\partial t}\int_\Sigma \Phi_1^2 ds=-\oint_C \mathfrak{v}\Phi_1^2 d\mathfrak{r}.$$

Aus den Beschränkung der zeitlichen Tendenz des Maßes der Störgrößen, als das das Flächenintegral intepretiert werden kann, ergibt sich die Integralbeziehung

$$\oint_C \mathfrak{v}\Phi_1^2 d\mathfrak{r}\geqq 0. \tag{4}$$

Für die verbleibenden zwei Gleichungen aus (2) ist das Auffinden einer Randintegralbeziehung durch die Schwierigkeit einer skalaren und einer vektoriellen Beziehung gekennzeichnet. Als Verknüpfung beider ergeben sich die folgenden zwei Gleichungen

$$\begin{aligned} &\frac{\partial}{\partial t}\left\{\sqrt{gH}\,(\Phi_2^2+\vec{\Phi}_3\vec{\Phi}_3)\right\}+\mathfrak{v}\sqrt{gH}\,\nabla\,(\Phi_2^2+\vec{\Phi}_3\vec{\Phi}_3)\\ &+gH\,\nabla(2\Phi_2\vec{\Phi}_3)=0\\ &\frac{\partial}{\partial t}\{2\Phi_2\vec{\Phi}_3\}+\mathfrak{v}\,\nabla(2\Phi_2\vec{\Phi}_3)+\sqrt{gH}\,\nabla(\Phi_2^2\\ &+\vec{\Phi}_3\vec{\Phi}_3)=0. \end{aligned} \tag{5}$$

Diese beiden Gleichungen ineinander eingesetzt liefern endlich die Beziehung

$$\begin{aligned} &\frac{\partial}{\partial t}\{\sqrt{gH}\,(\Phi_2^2+\vec{\Phi}_3\vec{\Phi}_3)-2\mathfrak{v}\Phi_2\vec{\Phi}_3\}\\ &+(gH-\mathfrak{v}\mathfrak{v})\,\nabla(2\Phi_2\vec{\Phi}_3)=0. \end{aligned} \tag{6}$$

Aus der Notwendigkeit einer Fallunterscheidung für die Interpretation des Flächenintegrals als Maß, ergibt sich die Bedingung

$$\oint_C (gH-\mathfrak{v}\mathfrak{v})\Phi_2\vec{\Phi}_3 d\mathfrak{r}=0.$$

Aus der Allgemeingültigkeit der Annahme, daß $gH\neq\mathfrak{v}\mathfrak{v}$ ist, folgt letztlich die Bedingung

$$\oint_C \Phi_2\vec{\Phi}_3 d\mathfrak{r}=0. \tag{7}$$

3. Ableitung der Randbedingungen

Auf Grund der Beziehung (4) und (7) lassen sich Aussagen bezüglich der vorzuschreibenden Randbedingungen machen. Zum besseren Verständnis sei folgende Annahme (siehe Abb. 1) gemacht. Die Strecke $\widetilde{AB}$ charakterisiere die Einströmsituation und die Strecke $\overline{AB}$ die des Ausströmens.

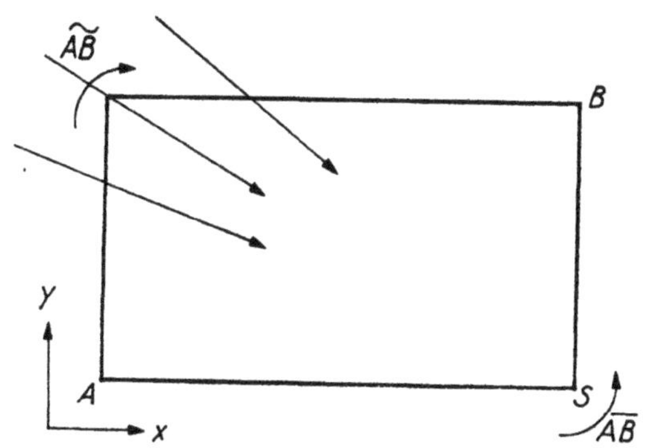

Abb. 1. Schematische Darstellung der Ein- und Ausströmsituation

Die Integralbeziehung (4) läßt sich demnach in zwei Teilintegrale aufspalten

$$\oint_C \mathfrak{v}\Phi_1^2 d\mathfrak{r} = \int_{\overline{AB}} \mathfrak{v}\Phi_1^2 g\mathfrak{r} - \int_{\widetilde{AB}} \mathfrak{v}\Phi_1^2 d \geqq \mathfrak{r}0. \tag{8}$$

Eine hinreichende Bedingung dafür, daß diese Beziehung erfüllt ist, ist durch

$$\int_{\overline{AB}} \mathfrak{v}\Phi_1^2 d\mathfrak{r} \geqq 0, \quad -\int_{\widetilde{AB}} \mathfrak{v}\Phi_1^2 d\mathfrak{r} \geqq 0$$

gegeben. Auf Grund der Abschätzung

$$\int_{\overline{AB}} \mathfrak{v}\Phi_1^2 d\mathfrak{r} = U \int_A^S \Phi_1^2 dx + V \int_S^B \Phi_1^2 dy \cong U \sum_A^S \Phi_1^2 \Delta x + V \sum_S^B \Phi_1^2 \Delta y = K > 0$$

folgt die Bestimmtheit des Integrals über $\overline{AB}$, hingegen ergibt sich bezüglich $\widetilde{AB}$ eine Bedingung etwa derart, daß $\Phi_1 = 0$ gesetzt wird.

Für die Integralbeziehung (7) läßt sich unter Aufspaltung in die beiden Teilintegrale auch die folgende Beziehung finden

$$\int_{\overline{AB}} \Phi_2 \vec{\Phi}_3 d\mathfrak{r} = \int_A^S \Phi_2 \frac{\partial h}{\partial x} dx + \int_S^B \Phi_2 \frac{\partial h}{\partial y} dy = \Phi_2 h \Big|_A^B - \left\{ \int_A^S h \frac{\partial \Phi_2}{\partial x} dx + \int_S^B h \frac{\partial \Phi_2}{\partial y} dy \right\}$$

$$\int_{\overline{AB}} \Phi_2 \vec{\Phi}_3 d\mathfrak{r} = \underset{\overline{AB}}{\Phi_2 h} - \int_{\overline{AB}} h \nabla \Phi_2 d\mathfrak{r}. \tag{9}$$

Unter Beachtung der hinreichenden Bedingungen dafür, daß das Integral (7) zu Null wird, ergibt sich für $\widetilde{AB}$ und $\overline{AB}$ die Notwendigkeit, eine Bedingung vorzuschreiben, etwa nach Beziehung (9) die Variable $h = 0$.

4. Schlußbemerkungen

Als Resultat der vorangegangenen Untersuchungen läßt sich folgendes formulieren:

An Einströmpunkten, durch $\widetilde{AB}$ charakterisiert, müssen zwei Randbedingungen, an Ausströmpunkten, durch $\overline{AB}$ dargestellt, muß hingegen nur eine Randbedingung vorgeschrieben werden. Eine Möglichkeit der Wahl der Randbedingungen ist die, stets das Geopotential vorzuschreiben und zusätzlich bei Einströmen noch die Vorticity, die ansonsten von innen heraus bestimmt ist.

Dieses Ergebnis deckt sich mit der in der Praxis durch heuristische Überlegungen gefundenen Wahl der Randbedingungen.

Literatur

Elvius, T., A. Sundström, Computationally efficient schemes and boundary conditions for a fine-mesh barotropic model based on the shallow-water equations. Tellus 25, (1973) S. 132

Herzog, H.-J., A. Meyer, Zur approximativen Ermittlung seitlicher Randbedingungen für die trockenen baroklinen primitiven Gleichungen anhand eines linearisierten Systems. Gerlands Beitr. Geophysik 85 (1976) 4, S. 283

Kreiss, H., J. Oliger, Methods for the approximate solution of time depended problems. GARP-Publications Series N 10 (1973), S. 107

Anschrift: Dipl.-Math. A. Meyer, Meteorologischer Dienst der DDR, Zentrale Wetterdienststelle, DDR-15 Potsdam, Michendorfer Chaussee 23

Comment on „Zur Anwendung der Korrekturmethode bei der numerischen Analyse mittels optimaler Interpolation".

By **Jan van Maanen** and **Seijo Kruizinga**

In a recent paper of Dr. J. Kluge (1) a statement is made about objective analysis with a combination of a correction method and optimum interpolation. In short, this statement is as follows.

Suppose we have observations of a variable $\tilde{f}_{ki}$, $k=1,\ldots,m$; $i=1,\ldots,n_k$. We want to make an estimate of the value of the variable f at some gridpoint:

$$(1)\qquad \tilde{f}_{00}=\sum_{k=1}^{m}\sum_{i=1}^{n_k} p_{ki}f_{ki}\,.$$

For details of the notation used we refer to the article of J. Kluge. If the weights p_{ki} are determinded according to the normal equations

$$(2)\qquad \sum_{l=1}^{m}\sum_{j=1}^{n_l}\overline{\tilde{f}_{lj}\tilde{f}_{ki}}\,p_{lj}=\overline{\tilde{f}_{ki}\tilde{f}_{00}},\quad k=1,\ldots,m;\quad i=1,\ldots,n_k$$

then the method of analysis is called optimum interpolation.

In the third section of the mentioned article the observations are used in two stages, $k=1,\ldots,m^I$ and $k=m^I+1,\ldots,m$.

First, $\tilde{f}^I_{00}$ is computed according to

$$(3)\qquad \tilde{f}^I_{00}=\sum_{k=1}^{m^I}\sum_{i=1}^{n_k} p^I_{ki}\tilde{f}_{ki}$$

in which the p^I_{ki} are not necessarily determined with optimum interpolation. Next, the remaining observations $k=m^I+1,\ldots,m$ are combined with $\tilde{f}^I_{00}$ according to

$$(4)\qquad \tilde{f}^{II}_{00}=p^{II}_{00}\tilde{f}^I_{00}+\sum_{k=m^I+1}^{m}\sum_{i=1}^{n_k} p^{II}_{ki}\tilde{f}_{ki}$$

and p^{II}_{00} and the p^{II}_{ki} are determined with optimum interpolation. The conclusions is that $\tilde{f}^{II}_{00}$ is equal to $\tilde{f}_{00}$ obtained in eq. (1), where all information was used in one step.

If this statement were true, then it would be possible to set constraints on a part of the interpolation weights and nevertheless reach the same optimum. This is clearly impossible for arbitrary constraints, so it is worthwhile to make a deeper analysis of the proof.

Eqs. (17) and (18) of the article of J. Kluge have the same right-hand side, and therefore it is true that

$$(5)\qquad \sum_{k=1}^{m^I}\sum_{i=1}^{n_k}\sum_{l=1}^{m^I}\sum_{j=1}^{n_l} p^{II}_{00}p^I_{lj}p^I_{ki}\overline{\tilde{f}_{lj}\tilde{f}_{ki}}+\sum_{l=m^I+1}^{m}\sum_{j=1}^{n_l}\sum_{k=1}^{m^I}\sum_{i=1}^{n_k} p^{II}_{lj}p^I_{ki}\overline{\tilde{f}_{lj}\tilde{f}_{ki}}=\sum_{l=1}^{m^I}\sum_{j=1}^{n_l}\sum_{k=1}^{m^I}\sum_{i=1}^{n_k} p_{lj}p^I_{ki}\overline{\tilde{f}_{lj}\tilde{f}_{ki}}$$
$$+\sum_{l=m^I+1}^{m}\sum_{j=1}^{n_l}\sum_{k=1}^{m^I}\sum_{i=1}^{n_k} p_{lj}p^I_{ki}\overline{\tilde{f}_{lj}\tilde{f}_{ki}}\,.$$

However, it is incorrect to conclude now that

$$(6)\qquad p^{II}_{00}p^I_{lj}=p_{lj},\quad l=1,\ldots,m^I,\qquad j=1,\ldots,n_l$$

and

$$p^{II}_{lj}=p_{lj},\quad l=m^I+1,\ldots,m,\quad j=1,\ldots,n_l,$$

which es eq. (19) in the article of J. Kluge. This conclusion is justified only if one can vary the expressions $\overline{\tilde{f}_{lj}\tilde{f}_{ki}}$ independently of their coefficients in (5). In this case this variation of $\overline{\tilde{f}_{lj}\tilde{f}_{ki}}$ is not permitted. The coefficients p^{II}_{00}, p^{II}_{ki}, p_{ki} are strongly dependent on $\overline{\tilde{f}_{lj}\tilde{f}_{ki}}$, because the p^{II}_{00}, p^{II}_{ki} are determined with the aid of normal equations of the following form

$$(7)\qquad p^{II}_{00}\overline{\tilde{f}^I_{00}\tilde{f}^I_{00}}+\sum_{k=m^I+1}^{m}\sum_{i=1}^{n_k} p^{II}_{ki}\overline{\tilde{f}^I_{00}\tilde{f}_{ki}}=\overline{f_{00}\tilde{f}^I_{00}}$$

$$p^{II}_{00}\overline{\tilde{f}^I_{00}\tilde{f}_{ki}}+\sum_{k=m^I+1}^{m}\sum_{j=1}^{n_k} p^{II}_{lj}\overline{\tilde{f}_{lj}\tilde{f}_{ki}}=\overline{f_{00}\tilde{f}_{ki}},\quad k=m^I+1,\ldots,m;\quad i=1,\ldots,n_k$$

and the p_{ki} are determined with the equations (5).

Reference.

(1) *J. Kluge*, Zur Anwendung der Korrekturmethode bei der numerischen Analyse mittels optimaler Interpolation, Z. Meteor. *26* Heft 1, 1976, S. 20–24.

Address: Jan van Maanen, Seijo Kruizinga, Royal Netherlands Meteorological Institute, De Bilt, The Netherlands.

Anmerkung zum Comment von J. v. Maanen und S. Kruizinga

Von **J. Kluge**

Der im obigen Comment gegen den Schluß von Gleichung (17) und (18) auf Gleichung (19) meines oben bezeichneten Artikels vorgebrachte Einwand ist stichhaltig. Damit bleibt die Frage offen, ob und in welchem Umfange bzw. unter welchen Bedingungen eine in mehreren Schritten ausgeführte optimale Interpolation *theoretisch* zu einem Informationsverlust bzw. zu einer Erhöhung des mittleren theoretischen Interpolationsfehlers führt.

Inwieweit wird dadurch der Kern des Artikels berührt, der durch zwei Aussagen bezeichnet wird; erstens, durch die Darlegung der Transformation der statistischen Struktur durch einen Interpolationsvorgang und zweitens durch Klarstellung der Bedeutung, die eine korrekte Beachtung der Transformation der statistischen Struktur bei der etappenweisen Interpolation hat?

Nach der Fertigstellung des Artikelmanuskriptes wurden numerische Experimente durchgeführt, um festzustellen, inwieweit sich die Resultate einer Einschritt- und Mehrschrittinterpolation *praktisch* unterscheiden. Anlaß dazu waren zwei Überlegungen. Einmal ist es nur näherungsweise möglich, die Transformation der statistischen Struktur zu modellieren. Zum anderen gibt es eine theoretische Abschätzung von *Gandin*, 1963, zum Problem der „fiktiven Stationen", die die Vermutung nahelegt, daß etappenweise Interpolation mit Informationsverlust verbunden sein kann. Nachträglich darf man auch den Comment als Begründung anführen.

Bei den numerischen Experimenten wurde im Rahmen einer numerischen Einflächenanalyse des Geopotentials der AT 500 einmal eine gemeinsame Analyse der aerologischen Information und von SIRS-Daten und im anderen Falle wurde zunächst eine optimale Interpolation allein der aerologischen Information durchgeführt, die im zweiten Schritt mit der SIRS-Information kombiniert wurde. Ohne einer geplanten Veröffentlichung vorzugreifen, seien hier vorab folgende Resultate mitgeteilt, die den erörterten Gegenstand betreffen.

1. Die Ergebnisse der Einschritt-Interpolation sind mit denen der Zweischritt-Interpolation nicht identisch.
2. Von entscheidender Bedeutung für die Resultate der Zweischritt-Interpolation ist die Beachtung der Transformation der statistischen Struktur durch die vorangehende Interpolation.
3. Wenn bei der Zweischritt-Interpolation die Transformation der statistischen Struktur nach den im Artikel angeführten Ansätzen modelliert wird, sind die Unterschiede zwischen Einschritt- und Zweischritt-Interpolation gering.

Ich möchte den Autoren für ihren kritischen Hinweis danken.

Literatur

Gandin, L. S., Ob"ektivnyj analiz meteorologičeskich polej. Leningrad, 1963, 285 S.

Anschrift: Dr. J. Kluge, Meteorologischer Dienst der DDR, Zentrale Wetterdienststelle, DDR–15 Potsdam, Michendorfer Chaussee 23

Originalmitteilung 551.510.533
551.557

Ein Vergleich der Ergebnisse von Ionosphärendriftmessungen im Langwellenbereich über Mitteleuropa und Ostsibirien im Winter 1975/76

Von **R. Schminder, E. S. Kazimirowsky, D. Kürschner** und **W. F. Petruchin**

Mit 3 Abbildungen

1.

Die bekannten winterlichen Stratosphärenerwärmungen, die im Extremfall (Januar 1963) zu einer hemisphärischen Umkehr des meridionalen Temperaturgradienten führen können, in der Mehrzahl der Fälle aber nur regionale Ausdehnung besitzen, haben im Bereich der Strato- und Mesosphäre zwangsläufig eine Richtungsumkehr im Zonalwind zur Folge. An dieser Richtungsumkehr und der Zunahme der Amplitude der in Höhen über 70 km bereits merklichen Gezeitenwinde läßt sich die Ausbreitung der Störungen bis in den Bereich der oberen Mesopausenregion (90–100 km) hinein nachweisen (z. B. *Schminder, Kürschner*, 1974; *Sprenger* u. a., 1974; *Schminder*, 1977). Sie scheint über planetare Wellen zu erfolgen (*Lauter, Schminder*, 1976).

In einzelnen Fällen wurde beobachtet, daß in der oberen Mesopausenregion gemessene typische winterliche Windfeldabweichungen dieser Art im Bereich der Stratosphäre von nur geringfügigen Abweichungen vom Normalverhalten begleitet waren (z. B. im Februar 1967 über Mitteleuropa), oder daß ausgeprägte Stratosphärenerwärmungen nur lokal zu Windfeldabweichungen in der oberen Mesopausenregion führten. Letzteres war im Dezember 1967 der Fall, als die am 13. 12. 67 zuerst über West Geirinish (57° 21′ N, 7° 22′ E) in 40 km Höhe festgestellte Erwärmung (*Johnson*, 1969) zwar nach den Messungen am Observatorium Collm in 95 km Höhe über Mitteleuropa ohne wesentliche Auswirkungen blieb, wohl aber in 90 km Höhe über College, Alaska, (*Hook*, 1970) das Zonalwindsystem vorübergehend zum Umschlagen brachte.

2.

Um den zeitlichen Ablauf und die Ausbreitung dieser Störungen, die zu den eindrucksvollsten Ereignissen im Erscheinungsbild des Temperatur- und Windregimes der winterlichen Hochatmosphäre gehören, von zwei hinreichend weit voneinander entfernten Beobachtungspunkten aus studieren zu können, wurde zwischen dem Geophysikalischen Observatorium Collm der Karl-Marx-Universität Leipzig (GOC) und dem Institut für Erdmagnetismus, Ionosphäre und Radiowellenausbreitung der Sibirischen Abteilung der Akademie der Wissenschaften der UdSSR in Irkutsk (SIBIZMIR) im Rahmen der KAPG (Kommission der Akademien der Wissenschaften sozialistischer Länder für die multilaterale Bearbeitung des komplexen Problems „Planetare geophysikalische Forschungen“) eine Vereinbarung getroffen, durch den ein abgestimmtes Meßprogramm in der in Frage kommenden Zeit von Mitte Dezember bis Ende Februar die ganzjährig und täglich am Observatorium Collm auf 185, 227 und 263 bzw. 272 kHz durchgeführten Ionosphärendriftmessungen durch adäquate Messungen auf 200 kHz an der Station Badary des SIBIZMIR zu ergänzen. Die Entfernung zwischen beiden Stationen beträgt etwa 5500 km.

Leider zählt der Winter 1975/76, in dem das gemeinsame Meßprogramm zum ersten Mal voll wirksam wurde, zu den wenigen der letzten Jahre, in denen nur sehr schwache und regional begrenzte Stratosphärenerwärmungen auftraten. Es wurde lediglich ein sogenanntes „minor event“ gemeldet. Das Warmgebiet überdeckte vom 5.–11. 1. 76 in mittleren und höheren Breiten ein großes Gebiet vom Atlantik über Europa bis nach NE-Sibirien, wobei es am Ende dieser Periode in zwei Zentren über Zentral-Sibirien und dem NW-Atlantik zerfiel. Es kam zu keiner Aufspaltung des Polarwirbels.

3.

Für einen detaillierten Vergleich der Meßergebnisse auf 200 kHz am SIBIZMIR und der dazu ausgewählten benachbarten Frequenz 185 kHz am GOC mußte zunächst sichergestellt werden, daß die aus sehr indirekten Messungen gewonnenen Werte keine wesentlichen meß- oder auswertetechnisch bedingten systematischen Abweichungen voneinander zeigen. Physikalisch signifikante Abweichungen im Normalverhalten sind bei der Großräumigkeit der hochatmosphärischen Windsysteme infolge gleicher Bezugshöhe (einschließlich gleicher tageszeitlicher Änderungen derselben) und gleicher geographischer Breite nicht zu erwarten. Die mittleren monatlichen Tagesgänge für die Zonal- und Meridionalkomponente des Windes im Januar 1976 (Abb. 1) lassen erkennen, daß beides in erster Näherung zutrifft.

Zum Nachweis der Auswirkungen der Stratosphärenerwärmung auf das Windfeld in der Bezugshöhe der 185 kHz- und der 200 kHz-Driftmessungen, der hier auf das Verhalten des zonalen Grundwindes v_0 beschränkt werden soll, ist es wünschenswert, nach Möglichkeit alle Tage einzeln einer Tagesanalyse zu unterziehen, um Grund- und Gezeitenwindfeld voneinander abzutrennen. Das ist allerdings aus verschiedenen Gründen, die ihre Ursache letztlich im zeitweisen Fehlen von für die Anwendung der Meßmethode optimalen Bedingungen bei der Ausbreitung der Raumwellen im Langwellenbereich haben (z. B. unzureichende Meßdichte bei zu langsamen Fading), nicht immer zu erreichen, so daß Lücken in der Darstellung meist nicht völlig vermieden werden können.

Die Abb. 2 zeigt nun, daß tatsächlich in der Zeit vom 5.–15. 1. 76, also in der Periode der Stratosphärenerwärmung, der zonale Grundwind seine Richtung dreimal umkehrt, wobei für zwei Wechsel eine bemerkenswerte Übereinstimmung zwischen den Ergebnissen auf

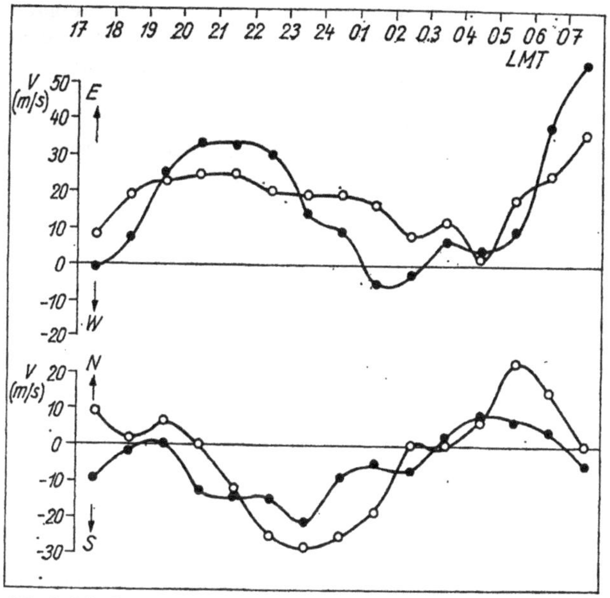

Abb. 1. Mittlerer monatlicher Tagesgang der gemessenen Zonal- und Meridionalkomponente des Windes in der oberen Mesopausenregion im Januar 1976 nach Messungen auf 185 kHz (Sender Zehlendorf/DDR) am GOC und auf 200 kHz (Sender Irkutsk/UdSSR) am SIBIZMIR. Bezugshöhe etwa 95 km, geogr. Breite bei 52°N, Entfernung der Meßpunkte etwa 5500 km, LMT = Ortszeit

●—● 185 kHz ○—○ 200 kHz

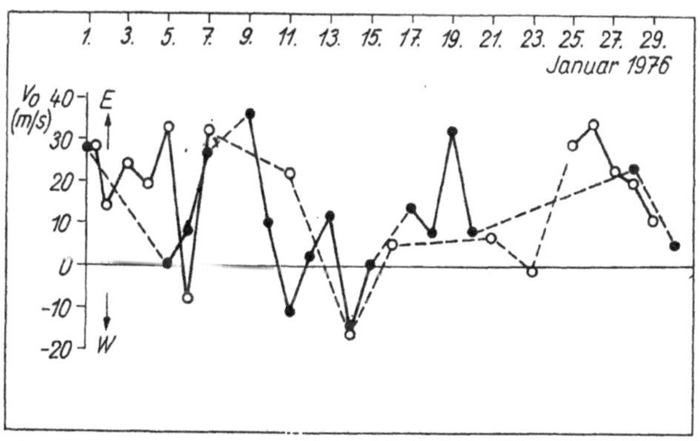

Abb. 2. Interdiurne Variationen des zonalen Grundwindes v_0(m/s) nach Messungen auf 185 kHz am GOC und auf 200 kHz am SIBIZMIR während eines „minor events" einer Stratosphärenerwärmung über Eurasien im Januar 1976. Über fehlende Tageswerte hinweg ist der Kurvenverlauf gestrichelt gezeichnet.

●—● 185 kHz ○—○ 200 kHz

beiden Meßstrecken vorhanden ist. Insbesondere ist die Zonalwindänderung vom 5. auf den 6. Januar nach den Messungen auf 200 kHz außerordentlich abrupt und zeitlich (um vermutlich einen Tag) gegenüber dem Auftreten auf der Meßstrecke auf 185 kHz verschoben. In der (hier nicht wiedergegebenen) Meriodionalkomponente wurden dagegen keine wesentlichen Abweichungen vom Normalverlauf gefunden.

4.

Der Vergleich der Meßergebnisse im Februar 1976 (Abb. 3) ergibt in der Zonalkomponente nur für die dritte Dekade eine hinreichende Übereinstimmung

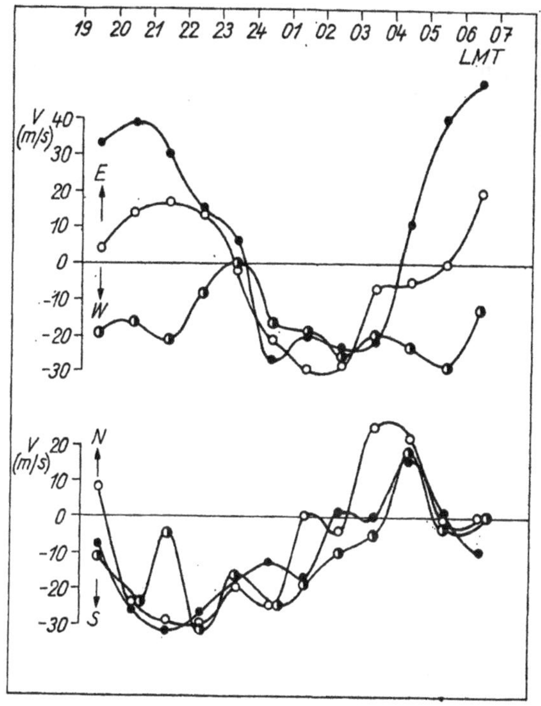

Abb. 3. Mittlerer monatlicher Tagesgang der gemessenen Zonal- und Meridionalkomponente des Windes in der oberen Mesopausenregion im Februar 1976 nach Messungen auf 185 kHz am GOC in Gegenüberstellung zu den Ergebnissen der Messungen auf 200 kHz am SIBIZMIR, bei denen nur die Werte vom 19.–29. 2. 76 mit denjenigen am GOC übereinstimmen, während vom 31. 1. –18. 2. 76 am SIBIZMIR eine Zirkulationsabweichung beobachtet wurde, wie sie im Zusammenhang mit Stratosphärenerwärmungen gefunden wird. LMT = Ortszeit

●—● 185 kHz, 1. 2.–29. 2. 76
◑—◑ 200 kHz, 31. 1.–18. 2. 76
○—○ 200 kHz, 19. 2.–29. 2. 76

der Werte. In der Zeit vom 31. 1.–18. 2. 76 wird in den Ergebnissen auf 200 kHz am SIBIZMIR eine starke Abweichung im Grundwind (und auch in den Gezeitenwindparametern) beobachtet, und zwar sowohl vom Normalverhalten als auch im Vergleich zu den Ergebnissen auf 185 kHz am GOC, die in diesem Fall weitgehend dem Normalverhalten entsprechen. Man kann sogar sagen, daß nach den Messungen am Observatorium Collm der Februar 1976 zu den relativ wenigen Februar-Monaten der letzten Jahre gehörte, die über Mitteleuropa eine ungestörte und dem Jahresgang entsprechende Zirkulation aufwiesen – mit schwachem zonalem Grundwind und großer Gezeitenwindamplitude, entsprechend der Lage im Minimum des Sonnenfleckenzyklus (*Sprenger, Schminder*, 1969).

Über Ostsibirien sehen die Verhältnisse dagegen offenbar anders aus. Hier kommt es in der Zeit, die am häufigsten von Zirkulationsstörungen im Zusammenhang mit Stratosphärenerwärmungen betroffen ist, nämlich in der ersten Februarhälfte, zu einer typischen Grundwindabweichung, die von derselben Art ist, wie sie bei Stratosphärenerwärmungen beobachtet wird.

Die Meridionalkomponente ist, wie Abb. 3 deutlich macht, praktisch wieder nicht betroffen.

5.

Es wäre ein großer Fortschritt, wenn es in der Zukunft gelänge, die vorhandene 5500 km-Trasse zu einem Meßpunktnetz mit einer Maschenweite in der Größenordnung von 1000 km auszubauen und gleichzeitig Meßwerte aus der Strato- und Mesosphäre aus diesem Raum operativ zur Verfügung zu haben. Das hier verwendete Meßverfahren der Ionosphärendriftmessung ermöglicht, beschränkt auf die Nachtstunden von Sonnenuntergang bis Sonnenaufgang, tägliche quasikontinuierliche Messungen, deren Ergebnisse bei Einsatz automatischer Meß- und Auswerteanlagen (*Kürschner*, 1975) unmittelbar zur Verarbeitung zur Verfügung stehen. Es ist daher in der Lage, zusammen mit dem apparativ etwas aufwendigeren Radarmeteorwindmeßverfahren für eine permanente Überwachung der Zirkulation der oberen Mesopausenregion (90–100 km) eingesetzt zu werden.

Literatur

Hook, J. L., Winds at the 75–110 km level at College, Alaska. Planet. Space Sci. 18 (1970), 1623–1838.

Johnson, K. W., A preliminary study of the stratospheric warming of December 1967–January 1968. Month. Weather Rev. 97 (1969), 553–564.

Kürschner, D., Konzeption und Realisierung eines vollautomatischen Registriersystems zur Durchführung von nach der D1-Methode angelegten Routinebeobachtungen ionosphärischer Driftparameter am Observatorium Collm. Z. Meteor. 25 (1975), 218–221.

Lauter, E. A., Schminder, R., Pressure and wind variations in the mesopause region, resulting from stratospheric warmings. Phys. Solariterr. 1 (1976), 111–119.

Schminder, R., Kürschner, D., Ergebnisse ionosphärischer Driftmessungen im 100 km-Niveau über Mitteleuropa im Sonnenfleckenmaximum 1968 bis 1970. Geophysik und Geologie, Geophys. Veröff. Karl-Marx-Univ. Leipzig, 3. Serie, Band I, Heft 1, 1974, 113–140.

Schminder, R., Abweichungen im Windfeld im Höhenbereich von 90–100 km über Mitteleuropa im Zusammenhang mit Stratosphärenerwärmungen. Gerlands Beitr. Geophys. 86 (1977), 105–114.

Sprenger, K., Schminder, R., Solar cycle dependence of winds in the lower ionosphere. J. atmos. terr. Phys. 31 (1969), 217–221.

Sprenger, K., Schminder, R., Greisiger, K. M., Kürschner, D., Schäning, B., Die Windsysteme in der oberen Mesopausenregion mittlerer Breiten nach Ionosphärendriftmessungen im Langwellenbereich. Institutsbericht. ZISTP(HHI) der AdW der DDR, 1974, 323 S.

Anschriften: Dr. rer. nat. habil. R. Schminder, Dr. rer. nat. D. Kürschner, Geophysikalisches Observatorium Collm der Karl-Marx-Universität Leipzig, DDR–7261 Collm
Dr. sc. nat. E. S. Kazimirowsky, Dr. rer. nat. W. F. Petruchin, Institut für Erdmagnetismus, Ionosphäre und Radiowellenausbreitung der Sibirischen Abteilung der Akademie der Wissenschaften der UdSSR, UdSSR–664697 Irkutsk 33, A/Ja 4

Hinweise für die Autoren

1. In der „Zeitschrift für Meteorologie" können Originalarbeiten (Übersichtsreferate, Aufsätze, kurze Originalmitteilungen, Buchbesprechungen, aktuelle Beiträge zur Chronik des Fachgebietes) in deutscher, russischer oder englischer, in Ausnahmefällen auch in französischer oder spanischer Sprache veröffentlicht werden. Es wird gebeten, den Titel der Arbeit in deutscher, russischer und englischer Sprache anzugeben.
2. Es wird darum gebeten, Manuskripte direkt einem der Herausgeber einzureichen. Die Herausgeber sind berechtigt, ggf. den Autoren Änderungswünsche zu unterbreiten. Anschriften: Prof. Dr. K. Bernhardt, Humboldt-Universität zu Berlin, Sektion Physik, Bereich Meteorologie und Geophysik, 1162 Berlin, Müggelseedamm 256/258. Prof. Dr. E. A. Lauter, Akademie der Wissenschaften der DDR, Zentralinstitut für solar-terrestrische Physik (Heinrich-Hertz Institut) 1199 Berlin, Rudower Chaussee 5. Prof. H. Pleiß, 809 Dresden, Am Kronenhügel 4
3. Jedem Aufsatz sollte vom Autor eine Zusammenfassung von maximal 8–10 Schreibmaschinenzeilen in deutscher und mindestens einer Fremdsprache (russisch, englisch) vorangestellt werden.
4. Die Manuskripte sind in Maschinenschrift $1^1/_2$-zeilig auf A-4 Bogen erwünscht. Bei der Anfertigung der Manuskripte sind die TGL 6710 Blatt 1 (Maschinengeschriebene Manuskripte für Bücher, Broschüren und Zeitschriften) und die TGL 10–098 (Formeln in Druckerzeugnissen) zu beachten.
5. Abbildungen sind in gut reproduzierbaren Vorlagen einzusenden. Die Beschriftung auf den Abbildungen soll mit Bleistift erfolgen. Der Maßstab für eine evtl. Verkleinerung oder Vergrößerung ist in der rechten oberen Ecke ebenfalls mit Bleistift anzugeben.
6. Sollen bestimmte Manuskriptteile kleiner als der übliche Text oder kursiv bzw. gesperrt gesetzt werden, so ist das Manuskript für diesen Teil auch $1^1/_2$-zeilig zu schreiben, aber seitlich mit einem senkrechten Strich zu versehen und mit „Kleindruck", „kursiv" oder „gesperrt" zu kennzeichnen.
7. Für die Literaturangaben wird folgende Verfahrensweise empfohlen. Im Text: Angabe des Verfassers und der Jahreszahl, bei mehreren Artikeln des gleichen Verfassers im gleichen Jahr sollte bei der Jahreszahl der Zusatz a), b,) c), usw. gemacht werden. Im Literaturverzeichnis: Name und abgekürzter Vorname des Autors, Titel der Arbeit, Angabe der Zeitschrift, des Bandes, des Erscheinungsjahres sowie der Seitenzahl. Bei Büchern sind die entsprechenden Angaben erwünscht unter Hinzufügung des Verlages, des Erscheinungsortes und bei Neu- und Nachauflagen der Auflagennummer.
8. Den Autoren wird empfohlen, nur die gesetzlich verbindlichen Maßeinheiten und deren Abkürzungen zu verwenden (Gesetzblatt der DDR vom 1. März 1969, Sonderdruck Nr. 605).
9. Die eingereichten Arbeiten dürfen an keiner anderen Stelle bereits veröffentlicht worden sein; die Autoren verpflichten sich zum Druck angenommene Arbeiten nicht anderweitig zu veröffentlichen.
10. Die Redaktion macht darauf aufmerksam, daß der Verlag berechtigt ist, dem Autor den über 10% der Gesamtkosten des Satzes hinausgehenden Betrag für Korrekturen, die nicht durch Verschulden der Druckerei entstanden sind, in Rechnung zu setzen.
11. Jeder Autor erhält 50 Sonderdrucke kostenlos; bei Arbeiten mehrerer Autoren erhält jeder Autor 30 Sonderdrucke.
12. Nachdrucke und Übersetzungen sind nur unter Angabe der Quelle und mit Genehmigung des Verlages gestattet.
13. Für den Inhalt der Arbeiten sind ausschließlich die Autoren verantwortlich. Die Redaktion ist für alle den Druck der Arbeiten betreffenden Fragen zuständig.
14. Die Anschrift der Redaktion lautet: Redaktion „Zeitschrift für Meteorologie", DDR 15 Potsdam, Kleine Weinmeisterstr. 14.

Inhaltsverzeichnis

Contents

Содержание

Bei Fragen zur Produktsicherheit wenden Sie sich bitte an:
If you have any questions regarding product safety,
please contact:

Walter de Gruyter GmbH
Genthiner Straße 13
10785 Berlin
productsafety@degruyterbrill.com